PRODUCT LIABILITY AND INNOVATION

Managing Risk in an Uncertain Environment

Janet R. Hunziker and Trevor O. Jones, Editors

NATIONAL ACADEMY OF ENGINEERING

NATIONAL ACADEMY PRESS
Washington, D.C.   1994

NATIONAL ACADEMY PRESS • 2101 Constitution Ave., NW • Washington, DC 20418

NOTICE: The National Academy of Engineering was established in 1964, under the charter of the National Academy of Sciences, as a parallel organization of outstanding engineers. It is autonomous in its administration and in the selection of its members, sharing with the National Academy of Sciences the responsibility for advising the federal government. The National Academy of Engineering also sponsors engineering programs aimed at meeting national needs, encourages education and research, and recognizes the superior achievement of engineers. Dr. Robert M. White is president of the National Academy of Engineering.

This publication has been reviewed by a group other than the authors according to procedures approved by a National Academy of Engineering report review process. The interpretations and conclusions expressed in this volume are those of the authors and are not presented as the views of the council, officers, or staff of the National Academy of Engineering.

Partial funding for the activity that led to this publication was provided by the Alfred P. Sloan Foundation and by Aetna Life and Casualty Company. Primary support was provided by the National Academy of Engineering Fund.

Library of Congress Cataloging-in-Publication Data

Product liability and innovation: managing risk in an uncertain
   environment / Janet R. Hunziker and Trevor O. Jones, editors.
     p.   cm.
   Includes bibliographical references and index.
   ISBN 0-309-05130-4
   1. Technological innovations. 2. Product liability. 3. Risk
management. I. Hunziker, Janet R., 1952– . II. Jones, Trevor O.,
1930– .
   T173.8.P7253   1994
   658.5'6—dc20   94-32031
                                            CIP

Printed in the United States of America

Steering Committee on
Product Liability and Innovation

Preface

————————————————

A t first glance, the realms of product liability law and the corporate research and development process seem worlds apart. Yet since the mid- to late-1980s, we had been hearing from our members, distinguished engineers from various fields, that it would be useful for the National Academy of Engineering (NAE) to explore the issue of product liability impacts on innovation. Product liability law is the realm of attorneys, but the effects of product liability law on the innovation process and especially on engineering development are of significant concern to engineers. The Academy has sought to bring to light the engineering consequences of liability law, by eliciting the views and experiences of engineers and others who have been deeply involved in the issue.

Formulating an activity to look at an issue where opinion tends to gather at the ends of the spectrum was not an easy task. On one end are those who point to record expenditures on R&D in some industries, noting that product liability obviously has not had an impact on investments in the process that leads to innovative products. On the other end are those who witness the reallocation of research dollars and the demise of complete product lines, attributing it mainly to the costs of the product liability system.

One of the first things the steering committee planning this activity recognized was that in many cases, the unpredictability of the system was a major problem for companies. Thus, the approach the committee decided to take was to look at how companies in some industries manage the risks and uncertainty inherent in designing, producing, and commercializing products and processes, given current trends in product liability law. The emphasis would be on best practice, with a particular focus on the perspectives and experiences of engineers. A symposium, "Product Liability and Innovation: Managing Risk in an Uncertain Environment," was organized and held September 20–21, 1993, in Washington, D.C., to examine these issues.

This volume is based on that symposium. The authors from industry provide insights into how the product liability system is affecting engineering functions as well as broader corporate practices. As these cases illustrate, just as impacts vary by industry, so too do the strategies for dealing with the product liability environment. The volume also gives perspectives on issues such as the efficiency of the product liability system, whether there is a causal effect between product liability law and safety innovations, what the legal and regulatory systems communicate about safety to product designers, the admission of scientific and technical evidence in the courtroom, and how knowledge of behavioral factors could be incorporated into product design to reduce risk.

I would like to thank Trevor Jones, who chaired the symposium and the Steering Committee on Product Liability and Innovation, and Janet Hunziker, the principal staff officer for the project, for their efforts in organizing the symposium and in the publication of this volume. Also, on behalf of the National Academy of Engineering, I would like to thank the committee members (listed on page *iii*) and the authors who participated in the symposium and submitted papers for this volume. A special note of appreciation goes to Robert Rines, chairman of the board of Franklin Pierce Law Center, for his summary of the proceedings at the symposium. Bruce Guile, director of the NAE Program Office, provided valuable guidance throughout the project. Thanks are also due other members of the NAE staff, including Dale Langford, Penny Gibbs, Maribeth Keitz, and Bette Janson for their able work on the symposium and the publication.

Partial funding for this activity has been provided by the Alfred P. Sloan Foundation and Aetna Life and Casualty Company. Primary support was provided by the National Academy of Engineering Fund.

Robert M. White
President
National Academy of Engineering

Contents

PRODUCT LIABILITY AND INNOVATION

Overview and Perspectives

TREVOR O. JONES AND JANET R. HUNZIKER

An engineer in the general aviation industry notes that in some cases 20 percent of engineering staff time is spent producing documents for various forms of legal discovery and preparing information for defense against product liability suits.

Engineers in the automotive industry, knowing that a design change can be misconstrued in a product liability suit to mean that the former design was deficient, feel constrained about discussing and implementing design changes, including safety improvements.

A manufacturer of life-saving implantable medical devices hears from some of its suppliers who are major producers of critical raw materials that they will restrict supply of their materials to that industry. The reason they give is that the risk of being pulled into product liability suits is too great. The company must fill the gap with materials from smaller, less technologically well-established companies.

As public policy debates go, the one involving the impacts of the U.S. product liability system on the ability of American companies to innovate and remain competitive may be perceived by most of the public as somewhat distanced from the normal activities of everyday life. Unlike the economy or health care, product liability is not an issue covered daily by the news media. Most Americans' knowledge of product liability comes from hearing or reading about particular cases or judgments that are deemed newsworthy because of the names of the corporate defendants, the circumstances of the case, or the size of the awards. Although tort reform is a recurring theme in federal and state legislation, most non-lawyers would be hard pressed to define what a "tort" is, let alone the reforms that are being recommended.

Yet, because the product liability system deals with the consequences of personal injury and suffering caused by the use of familiar, everyday products, it is a body of law that touches on universal experiences of human existence. Thus, people know in general terms why product liability law exists, namely, that it is intended to deter the manufacture and distribution of defective products, and when that fails, to compensate those who are injured by such products.

ROOTS OF THE DEBATE

Since the mid-1980s, however, there has been an increasing amount of debate in public policy circles about the effect of particular trends in product liability law. Even though product safety may have been improving, companies were experiencing more product liability cases and the size of the awards was increasing. As a result, their insurance costs were going up, and for some products, insurance coverage was being withdrawn altogether. More corporate resources were being expended on matters related to product liability, particularly in certain product areas. These trends were not limited to product liability, but extended to other types of personal injury cases, including medical malpractice, toxic substances, and accidents at public facilities (Committee for Economic Development, 1989). There followed a flurry of studies that sought to determine whether these impacts were measurable and could be documented. The studies included the following findings:

• There had been a fivefold increase in the number of product liability cases between 1975 and 1985, excluding asbestos-related claims (U.S. Department of Justice, 1987). The greatest growth has been in the area of mass torts where hundreds and even thousands of people claim to be injured by a single product (Committee for Economic Development, 1989; Hensler et al., 1987).[1]
• Although awards differ greatly depending on the case, personal injury awards had risen dramatically, particularly in the highest-paying cases. A study of two jurisdictions showed increases of 200 percent and 1,000 percent in the size of awards in constant dollars for product liability cases between 1960–1964 and 1980–1984 (Committee for Economic Development, 1989; Peterson, 1987).
• Insurance costs had risen dramatically in a period of a few years, in

some cases several hundred percent, for groups ranging from manufacturers of particular products to municipal governments (Committee for Economic Development, 1989; U.S. Department of Justice, 1986). This was in large part a response to the increase in the number of claims and the size of awards. Moreover, because premiums had failed to keep pace with losses through 1984, large increases over the next several years were needed to catch up (Harrington and Litan, 1988).

CONFLICTING EVIDENCE

Yet, for each contention that liability cases were indeed on the rise, and that the subsequent increase in litigation and litigation-prevention costs was throttling companies' ability to prosper, other data were cited to show that talk of a liability explosion and an overly burdensome tort system was unfounded. William Ide raises these points (in this volume), citing the low percentage that tort cases, and even more so, product liability, make up of the total state court caseload. On the specific issue of product liability impacts on innovation and competitiveness, he notes that the 1987 Conference Board survey of risk managers found product liability issues hardly affected larger economic issues, such as revenues, market share, or employee retention.

What becomes clear when listening to the arguments about this issue is that viewpoints and interpretations of data are dichotomized, with mythology on both sides, and getting an objective picture is extremely difficult. On the one hand, there is much talk of a litigation "crisis" in this country. Yet, on the other hand, studies show that one person in 10 who become accidentally injured turns to the tort system for compensation (Hensler et al., 1991) and that juries are becoming more pro-defendant in trial verdicts in personal injury cases (Reubi and Foster, 1994). Although a 1987 Conference Board survey of risk managers found product liability had a relatively insignificant impact on U.S. businesses, a 1988 Conference Board survey (McGuire, 1988) of 500 chief executive officers reached the opposite conclusion. Proponents of the current system say that because the product liability system adds on average less than 1 percent to the retail price of products, it cannot be the cause of America's competitiveness problems.[2] That may see like a small amount, but the percentage varies from industry to industry, and the total is still billions of dollars. This leads one to ask, What is the consumer getting from that surcharge, and where is that money going? While it is true that most states have laws that limit liability to what was known (state of the art) at the time the product was put into circulation, the practical reality of trial tactics today is that design changes come before juries all the time. Clearly, proponents of either side in this issue can find ample support for their arguments.

THE ENGINEERING PERSPECTIVE

The objective for this book, and the symposium that preceded it, was not to assess all the evidence about the direct and indirect effects of the U.S. product liability system. That is indeed a difficult issue to resolve. Parties at interest include numerous stakeholders—consumers, the legal community, engineers, insurers, corporate decision makers—involved in a debate over a range of products, which are used either out of desire or out of necessity and for whatever reasons are "targets" of product liability suits. Moreover, the diversity of stakeholders is compounded by the diversity of attitudes about risk and responsibility.

Rather, the goal was to get the engineering community's perspective on the effects of product liability on innovation. Why should we care about the experience of engineers? Because, as Richard Morrow points out, they are the *practitioners* of innovation. It is the engineers who think through the hundreds of details that result in the products we use everyday. What will the product do? What design will best accomplish that purpose? What materials will it be made of? How will different materials affect performance characteristics? How will the product be manufactured and packaged? If any group knows whether and how innovation is being affected by the product liability environment, it should be the engineers.

Moreover, it is difficult to talk about the engineering function in this context without also talking about (1) how broader corporate practice is affected by product liability, and (2) issues such as insurance, regulation, risk, and the presentation and interpretation of scientific and technical information in the courtroom. Indeed, the engineering function does not exist in isolation from the rest of an institution or society. Unlike the scientist, who often pursues research that may or may not have some eventual application, the engineer is motivated by the desire to produce a particular kind of product to fill a particular need. If regulations concerning that product change, it most likely will have an impact on research directions and product design and development. If obtaining insurance for a product becomes problematic, either because it is considered too risky for traditional insurance mechanisms or because self-insurance is not an option, that risk will have to be controlled somehow. This can be done through design changes or by eliminating the product completely. These messages come back to the engineer eventually, and affect how he or she does his or her job.

The unique value of this volume may be that it provides information on how some companies are responding to the risk they see presented by the product liability environment, highlighting the fact that there are costs, albeit varying, for the protection provided by our product liability system. These costs are manifested in various ways—in decisions made about where research dollars are allocated; in how engineers spend their time; in how products are manufactured, tested, marketed, and serviced; in choices about business opportunities; and eventually in the characteristics and availability of products for consumers.

THE PRODUCT LIABILITY–INNOVATION DYNAMIC

Three of the authors in this volume provide a background for the issue of product liability impacts on innovation: Richard Morrow, Victor Schwartz, and William Ide. Morrow describes why the intersection of product liability and innovation is so problematic, noting that we tend to take innovation for granted in the products we use, without understanding what it is. Innovation can be revolutionary or evolutionary, but in either case it implies improvement. It also implies risk, trade-offs, and making judgments about problems that could arise from product design changes. Engineers accept this because they know zero risk cannot be achieved. (On the issue of achieving zero risk, see also Breyer, 1993.) This ambiguity can be difficult for the law to handle, particularly when technical documentation of these trade-offs is interpreted by nonengineers or when nontechnical judges and juries are asked to pass judgment on highly technical decisions. The law, Morrow writes, tells engineers to make safe products, but it does not tell them how to be safe, or how safe to be. Morrow concludes that because each side sees so much at stake—the future of companies and classes of products, even American competitiveness, on the one hand, and individual safety and well-being on the other— resolution will not be easy. The debate causes us to reflect on what it means to live in America's litigious society in the late twentieth century, where technologically complex products precipitate battles over the issues of (1) corporate and personal responsibility and risk and (2) corporate and individual financial incentives.

Expansion of the System Creates Uncertainty

The roots of much of the current controversy over product liability can be traced to changes since the 1960s in the interpretation of the law or in the law itself that have expanded the product liability system. The shift from fault-based standards to strict liability has meant that manufacturers can be held liable even if a plaintiff does not prove negligence (see Schwartz in this volume). Products can be found defective even when they meet regulatory standards or conform to the state of knowledge at the time the product was produced. Expansion of the doctrine of joint and several liability allows plaintiffs to collect the full award amount from one defendant if the others cannot pay. Even the concept of contributory negligence has been relaxed, so that negligent plaintiffs can still recover damages. Rules concerning causation and statutes of limitation, particularly in long-latent injuries, have also been relaxed (Litan and Winston, 1988).

The cumulative effect of these changes is that it has become easier, in some cases, to bring product liability lawsuits against manufacturers. That is not to say that it has become easier for plaintiffs to win those cases, or that there are still not serious inhibitors that prevent most people from bringing such cases in the first place. Nevertheless, making it easier to bring lawsuits has meant that a manufacturer, whether in the right or not, has had to spend more of its resources, including those that would have been devoted to innovation, in defending itself, even if a case is settled.[3] As product liability rules were changing in the 1970s and early- to mid-1980s, it also became more difficult for a company to know what to do to prevent involvement in such suits.

Victor Schwartz notes at the beginning of his paper that a certain amount of stability has returned to product liability law, making product liability risk more manageable. In his description of the four ways a product can be found defective—manufacturing defects, innocent misrepresentation or express warranty, failure to warn, and design defects—he also notes what a manufacturer must do to avoid being found liable in any of those ways. Cases involving design defects or failure to warn, he writes, continue to be problematic.

The perceived unpredictability in the product liability system has prompted calls for reform at both the state and the federal level. The vast majority of tort cases are handled at the state level, 95 percent according to some sources (Hensler et al., 1987). States, however, have different product liability standards, and because product liability cases often involve litigants from different states, this can cause great uncertainty for manufacturers involved in interstate commerce. For years there have been attempts to develop some baseline federal standards for product liability law, but without success. Schwartz points out that a number of states have enacted their own reforms concerning punitive damages and design cases, but that to be truly effective, some fundamental issues such as joint and several liability, punitive damages, and the handling of claims must be dealt with at the federal level. (For a treatment of product liability reform issues, see Babcock, Appendix, in this volume.)

A frequently cited argument in support of federal tort reform is that foreign competitors have an edge because they do not have to deal with an unwieldy product liability system. Schwartz notes that foreign competitors have to follow the same rules as U.S. manufacturers when they sell products in this country. However, even if a foreign firm is subject to a U.S. court's jurisdiction, conducting depositions and collecting damages is much more difficult. Moreover, early penetration by U.S. companies of particular industries means they often have far more older products circulating in the marketplace than their foreign competitors. Because they are still liable for injury caused by those products, they have higher product liability costs, which are passed along to the consumer in higher product prices, making them less competitive on a price basis (Cortese and Blaner, 1989; Sontag, in this volume).

U.S. product liability law also differs in many ways from that of its major competitors. As Schwartz points out, these differences can be seen in measures related to the implementation of the European Community (EC) Directive as well as other aspects of the legal system. In Europe and Japan, for example, judges rather than juries decide cases, punitive damages are not awarded, and there are no contingent fees. More important, there are differences in the culture and the social attitudes toward risk, responsibility, and litigation as an avenue to redress wrongs. These cultural differences have created far different legal systems for dealing with product liability cases.

Another Perspective

The final background paper, that by William Ide, is strongly supportive of the American justice system, while noting the peculiar challenges of product liability law. His views represent another side of the debate and contradict the views of others in this volume. He notes that compelling reasons for having the current system are the number of accidents and deaths that occur when using a consumer product (although the product may not be the cause), the lack of a government-funded social safety net for accidental injuries, and the high cost of health care. Ide also discusses common misperceptions about the product liability system concerning numbers of cases, amounts of damages, and impact on U.S. innovation and competitiveness.

Ide does not contend that the current product liability system is perfect or that it should remain unchanged. He notes that legislatures and courts have been slow to respond to a number of challenges, including those brought on by advances in science and technology. As a result, the American Bar Association (ABA) has recommended changes that deal with issues such as uniformity of awards, punitive damages, and joint and several liability, and is working with other groups to develop a consensus on what improvements should be made and how they can be implemented.

Although many on the pro-reform side would criticize Ide's paper, it is important to understand the view he puts forth. It illustrates again that there are enough data and different interpretations of information (note in particular Schwartz and Ide's different views of the EC Directive) to support either side. His optimistic view of where the U.S. product liability system is headed is at odds with that of contributors from the chemical, medical device, and general aviation industries. In this, Ide differs with those who feel beleaguered by a product liability system that, in the hands of contingency fee-based lawyers and at times nontechnical judges and juries, seems neither wise nor reasonable.

IMPACT ON ENGINEERING PRACTICES, INNOVATION, AND CORPORATE STRATEGIES

Not all industries have been affected to the same degree by trends in product liability law. Authors in this volume look at five industries perceived as being heavily affected by product liability—chemical, general aviation, automotive, pharmaceutical, and commercial aviation. With the exception of general aviation, almost all Americans use products from these industries throughout their lives, and they are all industries where technological innovation is a critical component of business success and the continuing improvement of our quality of life. The writers of papers in this section are well qualified to comment on the impact of product liability on engineering practice and innovation, and the strategies companies employ to minimize the risk of product liability. They include engineers whose job it is to oversee research and development, and other high-level executives whose positions give them a broad view of how product liability is affecting their organizations. Although their comments are anecdotal, they indicate how product liability has affected different industries.

Even among these five industries, product liability has had varied impacts. It is blamed for the near decimation of an entire industry; it is considered the main determinant of what product lines a company pursues, thereby affecting supplies of materials for other critical industries; it is viewed as an expensive nuisance that diverts corporate resources but does not seriously restrict innovation.

Materials Suppliers

As described by Alexander MacLachlan, product liability has had a major impact on where R&D dollars are invested and which lines of business are pursued by DuPont, a highly diversified company in an industry that supplies a wide variety of important materials to other industries. He notes that because of the inherent dangers in manufacturing and handling chemicals, risk management and product stewardship are simply good business practices. Nevertheless, the risk of product liability litigation over products containing even a few cents' worth of DuPont materials has caused the company to rethink certain lines of business. The most dramatic example of this, and one that could potentially affect thousands of lives, concerns the provision of materials used in medical applications. Because of the high product liability risk, DuPont will stop supplying necessary materials to medical device manufacturers unless DuPont is involved in the design of the article and controls how the material is applied. An offshoot of this decision is that there will be little if any future investment by the company in research on synthetic material for internal-use medical devices.

Representing a company on the receiving end of this policy, Paul Citron is sympathetic to the dilemma faced by materials suppliers like DuPont. Because of their perceived deep pockets, these large companies become targets of lawsuits, even when they have no involvement in the specification, design, testing, or manufacture of the end product. He notes that defending against such lawsuits can cost millions of dollars on only dollars' worth of materials sales. With mature, technologically well-established firms being driven out of the materials supply business, the medical device industry has had to turn to smaller, less sophisticated materials manufacturers. Innovation in both the materials and the devices themselves has been slowed because it is in the exiting companies that breakthroughs would most likely have occurred, and the device manufacturers have had to divert resources to find new suppliers instead of advancing the state of the art.

General Aviation

If MacLachlan and Citron paint a bleak picture of how the product liability environment is affecting companies in the materials and medical device industries, the picture from the segment of the general aviation industry that produces piston engine-driven planes is even bleaker. Among the five industries represented in this volume, general aviation has been most affected by the product liability environment. This may be understandable considering the following circumstances: operators are not always proficient in handling the complex machines they fly, particularly in adverse weather conditions; aircraft maintenance may not always be thorough enough; and there are on average five, sometimes spectacular, accidents a day involving general aviation aircraft. The cost of defending against lawsuits, more than the cost of judgments, has had serious consequences. The light single- and twin-piston engine segment of general aviation is a beleaguered industry that of necessity innovates less and manufactures fewer new planes, creating a cycle in which there are more older aircraft and thus more accidents and product liability lawsuits.

Bruce Peterman's paper focuses on the engineering changes that product liability has brought about in the general aviation industry, noting that a few of these changes have been positive. Nevertheless, his description of the effects on various engineering functions, including allocation of resources, design, and documentation, reveals an industry in which engineers and technical personnel must consistently be in a defensive mode. Peterman implies that this diversion of the engineer's time to handling product liability-related concerns, whether it be direct involvement in a lawsuit or the overdesign of a product as a defensive measure, and the reluctance to include new technology in the product are a drain on engineering resources and are not conducive to innovation.

Frederick Sontag, as president of a company that supplies parts to the general aviation industry, provides a broader look at the indirect effects of product liability on a corporation in this industry. Among the papers in this volume, Sontag's gives the most comprehensive look at how product liability can make itself felt broadly in a company—from product strategies through relationships with other firms to financing. General aviation is arguably an industry in which the worst-case scenario of product liability impacts has occurred. Nevertheless, it is instructive to look at the unintended effects of a set of laws that were enacted to deter the manufacture of unsafe products and compensate victims of those products: restricted consumer choice in products and services, narrowed financing options, diversion of personnel and financial resources from productive activities, and higher product prices.

Automotive Industry

The impacts of the product liability system on the automotive industry, and the way that industry deals with product liability, are described in the papers by François Castaing and Charles Babcock. Castaing focuses on how product liability has affected engineering practice. He notes that the threat of lawsuits has had the unintended effect of (1) inhibiting the incentive to innovate, particularly safety features, (2) discouraging the critical evaluation of existing features, and (3) preventing the rapid implementation of improved design changes. He attributes these effects to trial tactics that misconstrue product improvements and design changes, particularly revolutionary ones, to mean that something about the former features was defective. He also examines the argument that if indeed product liability is responsible for safer cars, the United States should have the safest cars on the road; he notes that the Europeans and Japanese, which experience few product liability lawsuits, are known for their innovative and safe vehicles.

Consumer advocates may argue that the compulsion of law is required to motivate U.S. automakers (with their purported short-term perspective) to do what European and Japanese car manufacturers do without such coercion. Charles Babcock dispels this view in his paper that explores a range of issues—from the roots of popular assumptions about automakers and how highway safety could be improved through that industry's experience with product liability to the role of engineers in the debate. He asserts that product liability cases are nothing less than claims of engineering malpractice and that the U.S. legal system does not provide consistent messages about its rules for product safety and design to the audience that needs to hear them, namely, engineers. In dealing with engineers puzzled by how to respond to a product liability system that is perceived as capricious and unpredictable, Babcock suggests that the best advice to the engineer is simply to ask questions, innovate, and write accurate documents. On the question whether product liability discourages innovation in the automotive industry, Babcock cites evidence on both sides of the issue.

What then, he asks, is at the root of dissatisfaction with the U.S. product liability system? One factor may be a contrast between the popular wisdom that underpins this body of law, namely, that tort liability is an incentive for safer products, and the "reality" of the automotive industry, namely, that manufacturers make safer cars because the market demands it. Babcock also touches on doctrinal difficulties with product liability law, including whether it efficiently performs its functions of compensation and deterrence. He also notes that while overall highway safety is the fundamental problem addressed by this body of law, its focus on defects in new vehicles, which cause a statistically minute portion of highway accidents, has been misplaced. Despite declines in highway fatalities, which can be attributed to regulatory action, better highway design, and automotive safety devices, the leading cause of highway death and injury is driver behavioral factors, and it is this problem that must be addressed.

Commercial Aviation and Pharmaceuticals

If the papers in this volume are any indication, on a continuum indicating product liability impacts on innovation, general aviation is at one end, the chemical and automotive industries are in the middle, and the pharmaceutical and commercial aviation industries are at the other end. That is not to say that pharmaceuticals and commercial aviation are immune from the system, but rather that other factors—respectively, the regulatory system and the importance of public trust—are more important considerations.

Benjamin Cosgrove describes the product development cycle in the commercial aviation industry and how it responds to problems. Unlike the other industries described in this volume, however, commercial aviation sees product liability as neither a deterrent nor an incentive to innovation. The compelling reason for continuing to work toward safer aircraft is to maintain the company's reputation in an industry where public trust is everything. This is not to say that litigation costs are not high, that such costs are not a drain on resources, or that there is not a concern that the specter of liability might inhibit the ability of the company to innovate. Nevertheless, because reputation is so important, media coverage of aviation accidents is of greater concern than product liability.

Engineering practice and response to product liability in the pharmaceutical industry are influenced by the unique characteristics of the products produced, the way the products are marketed, and the fact that it is such a highly regulated industry. Marvin Jaffe describes the long, expensive process of gaining regulatory approval for drugs but notes that upon completing the process there is the assurance that the prescription drugs being put on the market are safe and effective, and that the risk of product liability has been lowered considerably as well. Nevertheless, Jaffe contends that there are some downsides for consumers from both the lengthy regulatory process and the tort system. These include negative impacts on pricing, orphan drug development, and patent protection, and constraints on innovation once compounds enter the regulatory pipeline. Product liability suits, while meant to keep harmful drugs and medical devices off the market, have also affected the availability and cost of particular critical pharmaceuticals, such as vaccines.

WHAT CAN BE LEARNED FROM THESE INDUSTRY CASES?

Product Liability Affects Industries Differently

A number of commonalities exist among the papers in this section. One is that the primary driver of safety innovations is not the fear of product liability lawsuits. Rather, most companies continue to innovate because it is just good business practice, it is necessary to maintain the public's trust in the industry, a global marketplace demands it, or because of regulatory mandate. Although several authors mention that demands for product safety have spurred innovation, they do not equate the product liability system with those demands.

The authors would also agree that the current product liability environment can act as a drag on a company by diverting critical engineering and financial resources from more productive activities. Beyond this, however, including product liability's role in inhibiting innovation, there is limited consensus among the industries. In most cases, it may be said that product liability does not inhibit innovation per se as much as it inhibits the introduction of some innovations into consumer products. Cosgrove, citing a corporate directive to continue innovating despite the prospect of litigation, indicates that the product liability system has a marginal impact on innovation in commercial aviation. In the pharmaceutical industry, Jaffe notes that a more serious constraint on innovation is the simple mechanics of the regulatory process and that liability's impact in this industry is felt more on particular categories of drugs. Castaing, giving the engineer's perspective, notes that the product liability system has made automotive engineers more cautious about evaluating and implementing design changes, with the result that innovation is avoided or slowed.[4] Babcock supports this charge, noting that the system does not give clear signals to engineers about how to design their products, but he also writes that given the global nature of the marketplace, it would be hard to name a design feature that is missing from U.S. automobiles because of product liability.

The most striking examples of unintended, negative effects of the product liability system, both on innovation and generalized business practice, come from the general aviation industry and the chemical industry. Peterman and Sontag, writing about particular segments of general aviation, show an industry on the defensive, where potential business agreements are not consummated, jobs are lost, and products are simply not produced because of product liability. MacLachlan and Citron write about major suppliers of materials used in critical medical device applications withdrawing from the market because of product liability. Here one sees substantial social and economic costs resulting from the product liability system. As Sontag notes, the exit from critical industries by major U.S. producers may create opportunities for overseas competitors, which may not be in the best interest of the consumer or the United States.

How Companies Manage the Risk of Product Liability

In any industry, product liability risk is part of the business environment, and as such it has to be managed like other risks. What do these papers tell us about how engineering practice has changed in response to product liability to lessen exposure to that risk? Again, this varies by industry. In almost all the industries it has meant greater attention to record keeping and more careful documentation of engineering decisions. The issue of overdesign is mentioned in the general aviation and automotive industry papers. It has also meant, in the chemical and pharmaceutical industries, pulling back from R&D investment in particular product areas.

At the corporate level, product liability cost control and litigation management techniques, indemnification, withdrawing from high-risk lines of business, and using technical expertise in the defense effort are all strategies for alleviating product liability risk. The papers suggest, however, that although these may be satisfactory strategies, they are not long-term solutions for dealing with product liability risk.

THE SOCIAL, LEGAL, AND REGULATORY ENVIRONMENT

The synergy between the product liability system and innovation takes place in a particular social, legal, and regulatory environment. The peculiar nature of that synergy is influenced by trends in such things as the availability of insurance, public attitudes toward risk, the body of statutory law that governs products and processes, and how technological questions are dealt with by the law. The last four papers in this volume examine these issues.

Insurance is one way manufacturers manage risk, and many observers contend that the unavailability of affordable insurance precipitated the "liability explosion." Leaving it to others to argue whether or not the insurance companies brought on their own problems in the 1980s, Dennis Connolly provides the insurer's perspective on how the insurance process works. He also describes the difficulty in assessing risk of technologically complex products, particularly those that are radical innovations and those, like many drugs, that have the potential to benefit large numbers of people but possibly harm a few as well. This perplexing underwriting task is exacerbated by particular product liability principles such as joint and several liability, strict liability, and the absence of caps on noneconomic damages, which complicates the assessment of risk, even for products from the most conscientious of companies. These and other problems that inject unpredictability into the system, Connolly notes, make it difficult for the insurer either to provide useful messages about risk to manufacturers, which would act as an incentive to improve loss experience, or to provide affordable insurance at all.

Peter Huber reiterates Connolly's points about how trends in product liability law have decreased the availability of intelligent, rational insurance. He focuses in particular on the problems that arise as a result of not knowing exactly what risks are being insured. The difficulty, he writes, is that as courts have become more accepting of marginal scientific theories, cause and effect has been trumpeted for risks far beyond those anticipated for a product. This lack of standards and tolerance of "junk science" in the courts, and the subsequent impact it has had on the availability of insurance, has had the most serious consequences for innovative technologies or products where rapid commercialization is most urgently needed. Huber sees the 1993 *Daubert* ruling, which dealt with the issue of admissibility of scientific evidence, as a step in the right direction, but describes other reforms that are also needed.

Susan Rose-Ackerman provides an important element to the discussion of the role of the regulatory system, or regulation by statute, in this debate. All five of the industries represented in this book are heavily regulated. Yet, as is well understood, complying with those regulations is not a guarantee that a firm will not be involved in product liability suits. What role, then, does the regulatory system perform, what incentives does it provide, and how does it interact with the regulatory effects of the tort system? Rose-Ackerman notes that different situations call for dependence on either torts or statutes, but that they can serve complementary roles as well. Regulatory reform through incentive-based regulatory statutes and solutions to the problem of providing compensation would, Rose-Ackerman contends, modify the judicial role and create a more efficient system, which could in turn, affect the research choices of firms.

Product liability law, insurance, and the regulatory system provide means for dealing with risk. Some people have argued that the expansion of the product liability system has resulted in a redistribution of economic wealth, and that it has forced companies into a social role for which they are not equipped, namely, insurers for any and all risk. The issue of risk is central to a discussion of product liability impacts on innovation for it is both an engineering and a social question: How much risk can be designed out of a product, and at what cost? How much risk should the user of a product be expected to assume? To what degree is a manufacturer responsible for injury that results from poor decision making by the user? Have we indeed reached the point where, as Norman Augustine (1994) observes, "Our system places a greater reward on assuring that nothing goes wrong than on assuring that something goes right"?

The paper by Baruch Fischhoff and Jon Merz examines how scientific knowledge about how people understand product risk can be incorporated into the product design and management process. Research in the areas of risk perception, judgment, and decision making provide insights into the ways people deal with complexity, new information, inconsistencies, and other factors that affect judgments. The authors describe how such knowledge could be applied by engineers and product designers to predict problems in the way products are used. Manufacturers could then improve warnings about potential risks as well as improve product design and the use of existing products. Unfortunately, Fischhoff and Merz note, the product liability system does not always provide incentives for manufacturers to consider behavioral issues. For example, the system acts as a disincentive if changing how risks are described on a warning label is construed as an admission that previous descriptions were inadequate. Means must be found for crediting manufacturers with incorporating behavioral issues into product stewardship.

WHAT CAN BE DONE?

Some may contend that the issue of impacts of a product liability explosion, particularly its effect on innovation and competitiveness, peaked in the late 1980s and early 1990s. Indeed, many studies cited in discussing this issue were done in the mid- to late-1980s and are becoming dated as we enter the mid-1990s. In this volume, Ide notes that product liability cases have actually declined in recent years, and Babcock cites the view of two legal scholars that there has been a pro-defense revolution in product liability law. Despite this, Babcock argues that recent experience in the automotive industry does not support the contention that product liability filings are declining. It is obvious too from other papers in this volume that product liability is still a serious concern for decision makers in some industries, particularly those producing certain categories of products. Numerous solutions that would alleviate some of the more burdensome aspects of the product liability system are put forth in the papers. These solutions range from more research in certain areas through specific engineering and corporate practices that would modify the risk of product liability to legislated reform of the product liability system.

Rose-Ackerman notes that although anecdotes are useful, they are incomplete as a basis for making policy. Thus, as Robert Rines pointed out in his summary remarks at the symposium on which this volume is based, there is a need for further documentation of decisions in a range of industries concerning innovation and quantification of the indirect and direct costs of product liability. Rines also suggested that more research needs to be done on whether there is a cause-and-effect relationship between product liability law and product safety. Babcock proposes a study of ways that legal systems of other countries treat "malpractice" by professional engineers, particularly ways that legal rules affecting engineering practice are communicated to the engineer.

Although such research would be useful from a public policy perspective, the authors also suggest numerous strategies to help engineers manage the risk of product liability on a day-to-day basis. Schwartz discusses the importance of paying attention to quality so that manufacturing defects are minimized, exercising caution in claims made about products, and accurately and effectively communicating risks. Careful documentation of decisions made during the product development process, although time consuming, is also important. Citron notes that in addition to producing high-quality products, firms must also track performance, expand understanding of first principles, and invest in improvements. Finally, Fischhoff and Merz make a strong argument for considering behavioral factors from the earliest stages of the product design process, which may contribute to product safety.

Although sweeping proposals for large-scale reform of the product liability system are frequently part of the debate on this issue, it is most likely that reform will be incremental. It will be done by individual judges and state legislatures, although Schwartz contends that some aspects of reform should be enacted at the federal level. Babcock, Citron, and Huber argue for reforms that would exempt manufacturers from liability, or at a minimum punitive damages, if they have conformed to safety standards. Rose-Ackerman also believes that a more enlightened regulatory system is in order and would alleviate some of the more problematic areas of the tort system.

Reforms that would affect the admissibility of scientific evidence were raised by several authors, including Huber and Jaffe. Huber and Citron urge the establishment of programs, similar to the national vaccine injury compensation program, for some niche products and services that have become "uninsurable." Several authors suggest that the inconsistency between laws concerning state of the art and the reality of the courtroom, where past engineering decisions are indeed judged by current standards, must be corrected. The costly and time-consuming nature of the discovery process was noted as one of the most onerous aspects of the current product liability system. Revised federal rules governing discovery, which became effective January 1, 1994, are intended to ameliorate that situation. At the state level, the American Bar Association has proposed a range of improvements concerning such things as uniformity of awards and excessive lawyer fees.

In addition to these specific reforms, certain issues about the product liability system may be resolved by no less than a national soul-searching about risk, particularly private risk, and responsibility, both individual and corporate. With every transaction that involves the purchase of goods, there is an implicit understanding between the producer and the consumer. The consumer assumes that the product is not faulty or unsafe, and the producer anticipates that the consumer will employ common sense when using it. Huber contends that consumer choice and fair warning do not count for much in the current system and that they should count for more. The second issue—who takes responsibility for injury—is broached by Babcock, who challenges the reader to consider the role of social insurance, specifically the effect that enactment of a national health care system might have on the compensatory function of the U.S. tort system.

Despite its benefits, the product liability system is perceived and experienced by many people, both plaintiffs and defendants, as complex, confusing, and unfair. Even though many cases can be cited that demonstrate that the system "works," the view that it does not work effectively for everyone and that it works inefficiently cannot be ignored and needs to be addressed. The ultimate irony may be that a body of law that was designed to reduce risk has in the end created more risk and uncertainty. Moreover, it treats alike both conscientious companies and those that knowingly commit acts that can cause harm. By altering the behavior of responsible companies, product liability law diminishes benefits to society.

Safe products and innovation are desirable goals that are in the public interest. The product liability system must ensure that they are not mutually exclusive.

NOTES

1. Examples of mass torts include the Dalkon Shield, with 325,000 cases, and asbestos, with over 60,000 cases settled, 100,000 cases awaiting action, and 2,000 more filed in court each month (Committee for Economic Development, 1989; Russakoff, 1994).

2. The median 1993 after-tax return on sales of the Fortune 500 is 2.9 percent (Fortune, April 18, 1994, p. 280).

3. One of the persistent uncertainties in trying to understand the costs and benefits of the U.S. product liability system is the costs borne by companies for legal representation whether a suit is settled or proceeds to a final conclusion. Since 90 to 95 percent of all civil cases are settled (see Ide, in this volume), it is likely that a considerable portion of a company's resources expended in product liability litigation are not related to jury awards.

4. In spite of regulatory pressures, some automotive braking systems and air bags were slowed in coming to market in the United States because of product liability concerns.

REFERENCES

Augustine, Norman R. 1994. Is any risk acceptable today? Across the Board 31(May):14–15.

Breyer, Stephen. 1993. Breaking the Vicious Circle: Toward Effective Risk Regulation. Cambridge, Mass.: Harvard University Press.

Committee for Economic Development. 1989. Who Should Be Liable? A Guide to Policy for Dealing with Risk. Washington, D.C.

Cortese, Alfred W., and Kathleen L. Blaner. 1989. The anti-competitive impact of U.S. product liability laws: Are foreign businesses beating us at our own game? The Journal of Law and Commerce 9(2):167–205.

Fortune. April 18, 1994. The Fortune 500: The Largest U.S. Industrial Corporations. 129(8):216–313.

Harrington, Scott, and Robert Litan. 1988. Causes of the liability insurance crisis. Science 239(February):737–741.

Hensler, Deborah R., Mary E. Vaiana, James S. Kakalik, and Mark A. Peterson. 1987. Trends in Tort Litigation: The Story Behind the Statistics. Report No. R-3583-ICJ. Santa Monica, Calif.: RAND.

Hensler, Deborah R., M. Susan Marquis, Allan F. Abrahamse, Sandra H. Berry, P. A. Ebener, E. G. Lewis, E. A. Lind, Robert J. MacCoun, Willard G. Manning, J. R. Rogowski, and Mary E. Vaiana. 1991. Compensation for Accidental Injuries in the United States. Report No. R-3999-HHS/ICJ. Santa Monica, Calif.: RAND.

Litan, Robert E., and Clifford Winston, eds. 1988. Liability: Perspectives and Policy. Washington, D.C.: Brookings Institution.

McGuire, E. Patrick. 1988. The Impact of Product Liability. Conference Board Report No. 908. New York: The Conference Board.

Peterson, Mark A. 1987. Civil Juries in the 1980s: Trends in Jury Trials and Verdicts in California and Cook County, Illinois. Report No. R-3466-ICJ. Santa Monica, Calif.: RAND.

Reubi, Marie, and Jill Foster. 1994. 1994 Current Award Trends. Horsham, Pa.: LRP Publications.

Russakoff, Dale. 1994. Asbestos pact: Legal model or monster? The Washington Post. May 11. A1, A12–A13.

U.S. Department of Justice. 1986. Report of the Tort Policy Working Group on the Causes, Extent, and Policy Implications of the Current Crisis in Insurance Availability and Affordability. Washington, D.C.: U.S. Government Printing Office.

U.S. Department of Justice. 1987. An Update on the Liability Crisis. A report of the Tort Policy Working Group. Washington, D.C.: U.S. Government Printing Office.

THE DYNAMICS OF
INNOVATION AND
PRODUCT LIABILITY

Technology Issues and Product Liability

RICHARD M. MORROW

———— ▬ ————

S ince earliest times mankind has created and used implements and devices to carry out the tasks of daily life. Hand tools, horsedrawn
plows, ladders, printing presses, and steam engines laid the foundation of today's world—a world that has not been particularly benign. For
example, in the ten-year period beginning in 1928 when reliable data first
became available, there were over 800,000 accidental deaths in the workplace, around the home, or on the highways (National Safety Council,
1992). Added to the fatalities were countless millions of injuries.

It is generally accepted that there has always been some degree of trade
off between the use of implements, machinery, or new technologies and
the personal risks inherent in their use. At various times accidents happened and people were injured because the devices were unsafe. Either
safe technology was not available or they were poorly designed, improperly manufactured, or became worn out and dangerous to use. On other
occasions accidents and injuries occurred because people used poor judgment or behaved in ways that are not always rational, such as drinking al-
cohol and then driving a car, removing a safety guard from a piece of industrial equipment, or knowingly diving into a shallow swimming pool.
Accidents also happen when products are improperly used beyond their
design limitation.

Attitudes about the amount of risk people should have to assume in using products and where to place the responsibility or blame when something goes wrong have changed over the past several decades. Years ago,
the social costs associated with physical injury were borne primarily by
the individual. Today, determining responsibility and seeking redress for

accidental injury or death have increasingly become contentious issues and are frequently resolved only through the litigation process. This is costly. Over the past 40 to 50 years, U.S. tort costs have risen significantly, reaching an estimated $132 billion in 1991 (Tillinghast, 1992). Such high costs are borne by everyone in one form or another, and often have consequences that are not planned or obvious. This is becoming an issue of growing national concern.

As providers of the goods and services we use in our daily lives, companies are viewed as both the cause of the problem and the answer to the question, "Who will pay for the physical and societal costs when an accident occurs?"

Product liability requires that companies critically examine the possible risks associated with their products or services. It also forces us, as a society, to look at the very personal nature of pain and suffering. Most of the time, companies go about their business of providing goods and services that the public wants. Usually when people get injured there is no connection between the two. This volume is about those times when there is an actual or alleged connection, the system of laws for dealing with it, and most important, the ramifications of that system.

DEFINING THE TERMS

Product liability and innovation. By themselves, these terms signal different disciplines. But in the following context they are frequently linked: Does product liability plus innovation equal a problem?

Product liability law has its roots in providing a means for those injured by defective products to seek redress. It aims not only to compensate victims but also to act as an incentive to providers of goods to make their products safe. The body of tort law that guides us today is of fairly recent development, having matured within the past 40 years. It is fair to say that the legal theorists who had much to do with the formation of modern tort law were well-intentioned and felt that society's best interests would be properly served by this framework for resolving disputes involving personal injury and accidents.

Innovation is the introduction of new methods or devices. By implication, one views these changes as improvements. Innovation can include a spectrum of change—from breakthrough discoveries in product and process to incremental improvements in design, materials, production methods, and quality control. Technological innovation has been an important contributor to the competitiveness of the U.S. manufacturing sector, to safer and more effective products, to job creation, and to the growth of this country's standard of living.

By definition, innovation involves risk. Change always does. No matter how much something is analyzed, tested, and evaluated, there is always the element of the experimental with an innovation. It is impossible to know exactly what the outcome will be. There are never any guarantees that an innovation will work, let alone work perfectly.

And what part does the engineer play in the innovation process? Engineers are "the practitioners of innovation." Although all of us, regardless of our professions, are practitioners of innovation in some form or other, it is those doing research and development, the engineers, who are at the cutting edge of technological innovation.

THE PRODUCT LIABILITY–INNOVATION LINK

We all know that laws are not static. As they change, there is a ripple effect on institutions and people that may not have been expressly anticipated or predicted. As mentioned earlier, it was anticipated, even intended, that changes in tort law, and product liability law in particular, would act as an incentive to bring better and safer products to market. However, we then must ask these questions: along with the good that product liability law has accomplished, have there been unintended deleterious effects? If so, what are they? How pervasive are they? Do they affect companies in any meaningful way?

An engineer working on a new product or process development, and his or her employer, cannot help but observe the following increasingly commonplace occurrences:

• Professional judgments concerning everything from design to end use can be introduced as evidence in product liability suits.
• Defendants can be held liable for products that were built according to accepted industry standards at the time; the plaintiff need only prove that it was possible to produce a safer product, even if the vastly higher production costs would have made the product virtually unmarketable.
• Manufacturers are increasingly being held responsible for human error and poor judgment in the use of their products.
• Largely nonscientific juries are being asked to make decisions on highly complex technical issues.
• Often the technology itself goes on trial, and the more unfamiliar the technology is, the more harshly it is judged.

The engineer may begin to feel that it is the courts that have the final say in what makes a good design or a good product.

The message our product liability system conveys to engineers is that they must design and produce safe products. What it does not tell them is how to be safe or how safe to be (Eads and Reuter, 1983). Although there has always been uncertainty in the introduction of new products and processes, many would contend that the impacts of product liability have added even greater uncertainty into every phase of planning for, and management of, the product cycle. This may be particularly true at the front end of the cycle, when decisions are made about whether to pursue development of a particular product and, if so, how it should be designed.

When innovation is stymied, products are less competitive, both here and in world markets where they must compete against products from countries, primarily Japan and the European Community, that do not bear the costs of a similar liability system. Moreover, the high costs of insuring against liability losses and defending suits funnels resources from productive activities and results in higher product prices. If indeed U.S. firms, in attempting to avoid exposure to liability lawsuits, are taking fewer risks and become less innovative, the ultimate loser is the consumer, the very one product liability was designed to protect.

This view of the issues is contradicted by those who say that there is a paucity of empirical evidence to support the contention that product liability deters innovation. While it may be true that the numbers—whether they be for punitive damages verdicts, award amounts, how many people actually bring lawsuits, or transaction costs—do not support the product liability–innovation link, it may also be true that numbers alone do not tell the whole story. This is a complex problem where perceptions of the legal environment, opportunities forgone, and innovations not pursued are a very real, albeit difficult-to-measure effect. Other intangibles that play a role in the experience a firm has with product liability are the degree of hazard inherent in the product, the ubiquitousness of the technology, and the number of people at risk. These variables affect both the degree of uncertainty in developing, manufacturing, and marketing a product, as well as the manner in which firms seek to lower their risk of exposure to liability.

SPECIFIC TECHNOLOGY ISSUES

The intersection of product liability and innovation raises a host of intriguing technology-related issues. The impact on design practice, the cornerstone of innovation, is perhaps the most salient. A recent National Research Council study (1991), noting the primacy of design, has stated that "quality cannot be manufactured or tested into a product, it must be designed into it." Since manufacturers can be held liable for a product design that exposes consumers to undue risk, it is incumbent on them to incorporate quality and safety measures into their product designs. This is no small task. Not only must industry and government standards be considered in the design, but today's products are increasingly complex, consisting of many different components and subsystems, often from different suppliers. These individual parts, as well as their interfaces, must be designed against failure and misuse.

This practice of "defensive design" incorporates a rigorous and careful application of engineering methods. It includes formal engineering analysis, testing, and anticipating problems through fault-mode and worst-case analysis methods. The objective is to explore not only where failures may occur but the implications of those failures. Design review, where a team of specialists examines every aspect of the design throughout the product life cycle, is crucial.

Written communication is an important part of this design process. Keeping records of data, processes, and the reasons particular decisions were made is common engineering practice. In the context of product liability, however, this written communication becomes problematic. Could this material, prepared with good intent and according to accepted practice, be subject to examination, and possibly used against the manufacturer in a court of law? If so, what is the solution?

Any engineer would agree that even with the most thorough testing and analysis of design, the most complete documentation, and the most carefully worded warnings and instructions, it is impossible to achieve zero risk. Moreover, there will always be people who misuse products, either accidentally or intentionally. There will always be trade-offs between the utility of the product and the danger that product may pose to the user. To cite an extreme illustration, knives, saws, and other cutting tools could be made with dull edges so that people would not injure themselves, but that would render them useless for their intended purpose.

In the real world, engineers are constantly making similar kinds of trade offs. A drug may successfully treat a life-threatening disease but also cause an allergic reaction in some patients. Stiffening the metal in auto bodies so that it absorbs more of the destructive force of a crash could actually shift more of the crash energy to the interior and the occupants of the car. Using environmentally safe chemicals in the manufacturing process may render mechanical components less safe if the chemicals do not provide an adequate level of corrosion resistance or structural integrity.

Clearly, expecting manufacturing institutions to insure against all risk is an untenable solution. The costs to our society—both financially through increased product costs and morally through the erasure of individual responsibility—would be too great to bear. What is needed is more dialogue about risk. This does not mean a one-sided communication from the "experts" about the nature of technological risk and its costs and benefits. Rather, it is necessary to create opportunities for all parties—individuals, groups, and institutions—to convey their values and concerns about products and their attendant risks (National Research Council, 1989).

Nowhere is this gulf between technical or scientific experts and the layperson more evident than in the courtroom. One of the strengths of the U.S. legal system is the guarantee of a trial by jury. But as products incorporate more complex technologies, it is imperative that both judges and juries be well-informed about the technological aspects of products and processes on which they are being asked to pass judgment. Moreover, there is now such a breadth of knowledge relevant to modern technological decisions that even members of the same technical or scientific disciplines disagree. This has given rise to the debate over the admissibility of scientific evidence, and the discretion judges have over what testimony is allowed at trial.

Finally, a discussion of product liability's impacts on innovation raises some basic questions about engineering and the law, and the way practitioners of those disciplines are trained to solve problems and view the world. Although facts dominate the worlds of both lawyers and engineers, the facts of lawyers are concrete, objective, and precise. The facts of engineers are data, signs, observations, and referents, meaningful only in relation to some organizing scheme (Nyhart and Carrow, 1983). In a study of the cognitive styles of lawyers and scientists, it was concluded that the structured, fact-based thinking of lawyers often conflicted with the structured, concept-based style of scientists and engineers. The study further points out that differences in cognitive style will influence "the effectiveness of communication, the degree of cooperation, understanding of questions, perception of truth, expertise and clarity of explanation" (Nyhart and Carrow, 1983, p. 236). It should be acknowledged that within each discipline there is a range of cognitive style. However, these generalized differences make it more difficult not only to establish trust and understanding across professional lines but also to translate scientific and technical information into a legal framework.

CONCLUSION

Although the topic of product liability impacts on innovation is one on which opinion tends to be extremely polarized, the one thing that most people would agree on is that there is a lot at stake. Some contend that the future of entire companies and classes of products, even the competitiveness of all U.S. industry, is at stake. Others insist that it is the well-being and safety of every American that is most at stake.

The resolution of this debate will not be easy, for it involves technological complexity, financial incentives, personal and corporate responsibility, risk, and other such issues that are often flashpoints for broader questions about what it means to live in America in the late twentieth century. Few of us would like to go back to the past when times were supposedly simpler, for they certainly were not any safer. What is needed is a critical ex-amination of where the path we are on is taking us, and whether we will want to be at that destination once we arrive there.

The papers in this volume, by providing the engineering view and iden-tifying important technological considerations concerning product liabil-ity and its impact on innovation, should add a new perspective to that as-sessment. Moreover, it is hoped that these papers will stimulate further analysis and study of this important topic.

REFERENCES

Eads, G., and P. Reuter. 1983. Designing Safer Products: Corporate Responses to Product Liability Law & Regulation. Santa Monica, Calif.: RAND Institute for Civil Justice.

National Research Council. 1989. Improving Risk Communication. Report of the Committee on Risk Perception and Communication, Commission on Behavioral and Social Sciences and Education and Commission on Physical Sciences, Mathematics, and Resources. Washington, D.C.: National Academy Press.

National Research Council. 1991. Improving Engineering Design: Designing for Competitive Advantage. Report of the Committee on Engineering Design Theory and Methodology, Manufacturing Studies Board. Washington, D.C.: National Academy Press.

National Safety Council. 1992. Accident Facts. Chicago: National Safety Council.

Nyhart, J. D., and M. Carrow. 1983. Law and Science in Collaboration: Resolving Regulatory Issues of Science-Technology. Lexington, Mass.: Lexington Books.

Tillinghast. 1992. Tort Cost Trends: An International Perspective. New York: Tillinghast.

Making Product Liability Work for You: A Path Out of the Product Liability Jungle

VICTOR E. SCHWARTZ

Many people view product liability law as a confusing morass of ever-changing rules. Tremendous uncertainties did mark the 1980s. Beginning in 1988, however, important stability returned to most of product liability law. Judges began to appreciate that expanding the liability system too far toward the plaintiff brings about adverse social consequences, such as the withdrawal of good products from the market and disincentives to innovation. One can use this new stability to help manage product liability risk.

To appreciate what stability there is and is not, one should focus on the fact that there are basically four ways to find a product defective: manufacturing defects, innocent misrepresentation or express warranty, design defects, and failure to warn.

MANUFACTURING DEFECTS

If a company makes a product, and the product is different from its own specifications and hurts somebody, the company is liable. For example, if a jar of peanut butter has a piece of glass in it, that is a manufacturing defect. If a car's torsion bar does not perform within specified limits, that is also a manufacturing defect.

In this area, strict liability, which means that liability is imposed even though the manufacturer was not in any way at fault and which began in the United States in 1963,[1] is really strict. There are some defenses, but they are predicated on rather unusual examples of user conduct, for example, total misuse of a product, such as driving a car across a river. This creates a very strong argument for the importance of quality control, since break-downs in quality control result in nearly inescapable liability for the man-ufacturers.

INNOCENT MISREPRESENTATION

A second way a product can be found defective is through innocent mis-representation or express warranty, in other words, what the manufac-turer says about the product.[2] If promises made about the safety of a prod-uct turn out to be untrue, even if they are not printed, the manufacturer can be held liable. A classic case happened in 1930 when an automobile company said that its glass would not shatter.[3] The glass was hit extraor-dinarily hard by a rock at high speed, and the glass shattered and hit the driver. The company was liable for the injury.

Two subtle changes have occurred in this area since 1988. First, the law used to be universal that the person bringing the suit had to hear or see the words expressing the promise, and then believing them to be true, relied on them. This is called the reliance factor. The person who was injured by the glass had to see an advertisement from the automobile company that made him that promise. In many states that is no longer true, and it is enough that the representation was made.

That has real importance, especially in workplace liability. When a man-ufacturer sells a product to another company, it may make certain claims about a safety aspect of that product. If the employee of the company that purchased the product is injured, he can bring an action against the man-ufacturer, even though the promises of safety were made to his employer, not directly to him. In the past, the employee could not bring an action for this because the manufacturer had not talked directly to him, but now that has changed in some states.

This issue of liability resulting from making promises to one group about a product being used by another group is also tremendously im-portant in the pharmaceutical industry. Pharmaceutical companies gener-ally do not talk directly to patients; they talk to doctors or hospitals. They make promises about their products, sometimes significant ones, though not to the user.

The second change in the interpretation of express warranty concerns the specificity of what is said about the product.[4] Formerly, the promises had to be very specific. There is a murkiness in the cases now so that state-ments that used to be regarded as simple exaggeration for promotional purposes—"This product is better than all the others" or "We have the safest one on the market"—may be used against the manufacturer in some jurisdictions.[5]

There are virtually no good defenses against a good express warranty claim since the user presumed that the product could do what was promised.[6] Neither the fault of the plaintiff, nor the fact that it was not scientifically possible to accomplish what was promised, is a defense. Tort reform will not change this other than to try to restore the reliance factor or to ensure that the promise has to be very specific. Because it is an area in which an increasing number of cases can be brought and won, it is important. The solution for the manufacturer is simply not to make the claims, either orally or in product literature or advertisements.

FAILURE TO WARN

The third way a product can be found defective is in failure to warn. Experience has shown that this is an area where manufacturers are vulnerable. This is because no matter how explicit the warning is, it can always be said that there should have been additional warnings. Cases have also been brought against manufacturers who put warnings only in English where that was not the first language of many of the product's users.

There has been some tightening of the rules in this key area of liability. In the late 1980s, there were cases that suggested a manufacturer would have to warn about risks even if it had no knowledge that those risks could occur.[7] In a 1991 case, however, the Supreme Court of California said that the state of the art, in the sense of what was knowable or could have been discovered, *was* a defense in a warning case.[8] Legislatures in New Jersey, Maryland, and Louisiana have also corrected cases that have extended liability beyond what could have been known and closed off what could have become open-ended liability.[9] In most states today, the law is that the state of the art is a defense in warning cases.

There are several things to keep in mind when writing warnings. The first rule is to get people's attention. Lettering, colors, and wording all come into play here. Second, in language that anybody can understand and in explicit terms, manufacturers have to explain the risk. If misusing the product is fatal, the user must be told. Third, and this is left out of many warnings, instructions must tell the user how to avoid the risk. With most products, there are ways to avoid risks and injuries, and if this is communicated well, the manufacturer should not be held liable.

Another cause of action under which liability has occasionally been imposed has been the "continuing duty to warn."[10] This means if after a product is made the manufacturer discovers a new and significant risk, the courts say there is a duty upon the manufacturer to take reasonable steps to warn the owners. The warranty card that comes with most consumer products allows manufacturers to keep track of who owns that product. If something goes seriously wrong with the product, the manufacturer should take reasonable steps to inform people about these newly discovered risks.[11] If the manufacturer no longer makes the product, it does not have a duty to discover risks. Courts have mandated recall for very few products, mainly aircraft parts.[12] So far, recall has been left largely to the regulatory agencies whose responsibility it is to oversee particular products.[13]

DESIGN DEFECTS

Design defects are the fourth way a company can be held liable for its products. To win these cases, the plaintiff must show the jury, in an understandable way, that there was a safer way to make the product. The Dalkon Shield litigation is an example of a case in which the plaintiffs' lawyers showed exactly what happened so that everybody on the jury understood it, and they demonstrated a feasible alternative design. If indeed there is a practical, economically feasible alternative design that would have increased safety, but it was not used, the manufacturer is going to be in trouble. In considering whether the alternative design is feasible, it is important to show that it would not bring about other, more serious risks.

TORT REFORM INITIATIVES

State Tort Reform

State tort reform efforts are under way to make both warnings law and design law more explicit so that the rules become more understandable and fairer. The work is being conducted by the American Tort Reform Association (ATRA) and the American Legislative Exchange Council (ALEC), both based in Washington, D.C.

Why is legislation needed now, when it was not needed in the past? Fifty years ago, courts were very serious about following *precedent*; however, over the past 25 years judges have engaged more in lawmaking, feeling it is their social duty to expand the rights of plaintiffs. This can have major impacts, for when the courts make a rule, it is retroactive, unlike rules that are developed by legislation.

In 1992 three states that are very different in their politics and geography—North Dakota, Mississippi, and Texas—have had successful tort reform efforts in the product liability area. In North Dakota the new law focused on punitive damages and limited them to two times the amount of compensatories or $250,000, whichever is greater.[14] In Mississippi the law outlined the four basic ways product liability can be brought (as set forth in this paper). The law also changed punitive damage rules by raising the burden of proof to "clear and convincing" evidence and permitting a defendant to have his punitive damage trial heard only if compensatory damages were awarded first.[15] In Texas the principal reform was to require that plaintiffs in design cases prove that there was an alternative way to make the product.[16]

However, there is at least one jurisdiction where a court has reverted to a 1980s style of open-ended liability. In 1992 the Supreme Judicial Court of Massachusetts said in a footnote that manufacturers can be subject to liability regardless of whether they knew or could have known about a risk.[17] In the past five years, when courts have gone as far as the Massachusetts judicial court, they have either made a self-correction and restored a "state-of-the-art" defense, or the legislature has overruled the decision.[18] If this Massachusetts case is not corrected, it could be exported to other states.[19] The law will dull innovation. It also may create chaos in the insurance industry, for if one has no idea what the risks are, how can they be insured against?

Federal Tort Reform

Tort reform at the federal level seems like it has been around as long as the pyramids, but unlike the pyramids it has changed. It has focused on product liability because products flow in interstate commerce. Enacting product liability tort reform at the federal level is difficult because those who are opposed to it are very strong politically in Washington. The product liability bills (S. 687 and H.R. 1910) of the 103rd Congress focus on punitive damages and joint liability. The Senate bill also includes reforms that would expedite claims and help injured persons with a discovery statute of limitations that would preserve claims until two years after a person discovered the injury and its cause.

One may ask which reforms will occur at the state level and which at the federal level. At this point, it seems likely that reforms dealing with design and the duty to warn will arise out of state tort reform. Overall reform of joint liability and punitive damages will occur at the state level, but to be totally effective it must occur at the federal level. There is new impetus and momentum behind the federal product liability bill; reform could come about within the next two years.

INTERNATIONAL COMPARISON

People often say that product liability law hurts the United States competitively. The quick response to that assertion is that a Japanese company selling a machine tool here is subject to the same liability laws as a U.S. company. One the other hand, U.S. companies tend to have more older products in the U.S. marketplace than their foreign competitors. Since in most states a company can be liable for a product for an extraordinary length of time, those U.S. manufacturers face higher product liability costs, which are passed on in the price of goods. United States machine tools have been here more than 100 years; the Japanese and German versions have been here for considerably less time.

The Europeans tried to modernize their product liability law beginning with the European Community (EC) Directive, which began to be developed in 1979 and was published in 1985. There are some key areas in which their law differs from U.S. law. One is that they use a single definition of "defect." Under the EC Directive, "a product is defective when it does not provide the safety which a person is entitled to expect, taking all circumstances into account." This definition is very open-ended and provides much room for argumentation.

There are helpful defenses in the EC Directive. First, if a defect is due to compliance of the product with mandatory regulations issued by public authorities, there is no liability. That is not the law in most of the United States. For example, the fact that a company followed National Highway Traffic Safety Administration regulations can be introduced as evidence, but is not a defense. In Europe it is a defense, and there is much to be said for that if the agencies are doing their jobs. How can a jury of 12 second-guess what an agency has taken 12 years to decide?

Second, most countries that have adopted the EC Directive limit liability to known technical knowledge that was in existence at the time the product was put in circulation. In the United States, that "state-of-the-art" limitation is the law in every state except Massachusetts and Hawaii. It is not mandatory in the EC Directive; for example, Luxembourg has derogated from this defense, as have Germany and Spain with respect to pharmaceutical products only. However, most countries have followed this limitation on liability.

Third, a mandatory 10-year statute of repose for all goods begins on the date the manufacturer put the product into circulation, unless proceedings had already been instituted by the injured party. Although Texas passed a 15-year statute of repose for capital goods in 1992, most states have no repose period. Cases can be found involving products more than 90 years old!

There are other key differences between the European liability system and that of the United States. In Europe, damages for pain and suffering are limited; generally in the United States they are not. The United States awards punitive damages; most European countries do not. In Europe, judges rather than juries decide cases, and there are no contingent fees (lawyers are paid by the hour).

CONCLUSION

The overall trends in American product liability law have been toward making it more rational. Some respected observers have called this trend a "quiet revolution."[20] Although the revolution is not complete, the changes have made the job of those seeking to manage product liability more feasible. As companies and engineers work toward eliminating the potential for product liability lawsuits, the law itself may continue to embody more rational thought and reasoning. Although this may be more a hope than a reality at present, it is a goal worth working toward.

NOTES

1. See *Greenman v. Yuba Power Prods.*, 377 P.2d 897 (Cal. 1963).
2. See Restatement (Second) of Torts § 402B (1965).
3. See *Baxter v. Ford Motor Co.*, 12 P.2d 409 (Wash.), *aff'd*, 15 P.2d 1118 (Wash. 1932).
4. See e.g., *Berkebile v. Brantly Heliocopter Corp.*, 311 A.2d 140, 146 (Pa. Super. 1973), *aff'd*, 337 A.2d 893 (PA. 1975); *Baughn v. Honda Motor Co., Ltd.*, 727, P.2d 655, 668 (Wash. 1986).
5. See, generally, *Crothers v. Cohen* 384 N.W.2d 562, 566 (Minn. App. 1986).
6. See Victor Schwartz, *Violation of Express Warranty: A Useful Tort That Must Be Kept Within Rational Boundaries*, 3 Prod. Liab. L.J. 147 (1992).
7. See *Beshada v. Johns-Manville Prods. Corp.*, 447 A.2d 539 (N.J. 1982); *Halphen v. Johns-Manville Sales Corp.*, 484 So. 2d 110 (La. 1986).
8. *Anderson v. Owens-Corning Fiberglas Corp.*, 810 P.2d 549 (1991).
9. See N.J. Stat. Ann. § 2A-58C-3 (West 1991); Md. Code Ann. Art. 27 § 36-I(h) (West Supp. 1990); La. Rev. Stat. Ann. §§ 9:2800.53, :2800.56.
10. See *Comstock v. General Motors Corp.*, 99 N.W. 2d 627 (Mich. 1959).
11. See *MC&C, Inc. v. Zenobia*, 587 A.2d 531, 544-45 (Md. Ct. Spec. App. 1991), *vacated on other grounds sub nom. Owens-Illinois v. Zenobia*, 601 A.2d 633 (Md.), *reconsideration denied*, 602 A.2d 1182 (Md. 1992).
12. See *Bell Helicopter Co. v. Bradshaw*, 594 S.W.2d 519 (Tex. App. 1979); *Braniff Airways, Inc. v. Curtiss-Wright Airways, Corp.*, 411 F.2d 451 (2d Cir.), *cert. denied*, 396 U.S. 959 (1969); *Noel v. United Aircraft*, 342 F.2d 232 (3rd Cir. 1964).
13. See *Smith v. Firestone Tire & Rubber Co.*, 755 F.2d 129 (8th Cir. 1985); *National Women's Health Network, Inc., v. A.H. Robbins Co.*, 545 F. Supp. 1177 (D. Mass. 1982).
14. N.D. Cert. Code § 32-03.2-11(4) (signed by governor Mar. 31, 1993).
15. Miss. H.B. 1270 § 2(1)(a) (signed by governor Feb. 18, 1993).
16. Tex. S.B. 4 (signed by governor Mar. 4, 1993).
17. *Simmons v. Monarch Mach. Tool Co., Inc.*, 596 N.E. 2d 318 (Mass. 1992).
18. See Victor Schwartz, *The Death of Super Strict Liability: Common Sense Returns to Tort Law*, Gonzaga L. Rev. 179 (1991/92).
19. A bill to overrule *Simmons* is pending in the Massachusetts legislature.
20. See James Henderson, Jr., and Theodore Eisenberg, *The Quiet Revolution in Product Liability*, 37 U.C.L.A. L. Rev. 479 (1990).

The Role of the Justice System in the Product Liability Debate

R. WILLIAM IDE III

xcellence in engineering in the United States has led to the design, manufacture, and sale to the world of the most innovative, efficient, durable—and safest—products on the market. It also made this country the world's leading manufacturing economy, creating unprecedented prosperity for its citizens. America's high-priest engineer, Buckminster Fuller, was an architect, a philosopher-poet, and an almost mystical optimist. With his crotchety single-mindedness, Fuller once said, "there is absolutely nothing that cannot be done." He also believed that "man can create miracles." The work engineers do on a daily basis is an affirmation of that belief in man's abilities.[1]

Our forefathers, when they created the American system, recognized the role freedom plays in unleashing this economic and social creativity. But they also saw another important requirement, and they created a justice system that has helped preserve our freedoms for more than 200 years. Few Americans would trade their justice system for that of any other country. By the same token, other nations have looked to the American justice system as a model for their own. This can be seen most recently in the way former Soviet bloc countries have studied and replicated many aspects of our justice system.

The fall of the Eastern bloc demonstrates that people cannot efficiently design and produce goods with strong government controls and directives such as those that were enforced by the Communist system. The genius of a capitalistic system is that it allows individuals to pursue their self-interests with minimal government involvement.

How is this accomplished in an orderly and consistent way, without some taking advantage of others? It is done through the rule of law. The United States is a free society, and neither government nor vested interest groups can infringe on certain basic personal freedoms. It is the right of every citizen to protect his or her legitimate interests within an independent judicial system. How to determine the legitimate interests of citizens and how to balance them against the concerns of others is what the law is all about. Nowhere is this more challenging for legislatures and courts than in the area of product liability.

The free market system and capitalistic ethic foster the continued growth of innovative product development in this country. But product development is dependent on consumers who are secure in the belief that they will be helped if injured through no fault of their own by a defective product. When government regulations or industry self-regulation fail, it is necessary to have a system that can identify and correct these products.

In the United States, the product liability system has been developed, through statute and case law, to compensate injured consumers and deter harmful products. Statistics on accidents demonstrate why this is important. Although the consumer product may not have been the cause, consumer products are involved in an estimated 29,000 deaths—more than die each year from such diseases as prostate cancer or emphysema—and 33 million injuries in the United States every year.[2]

In other industrialized countries, those who are injured generally receive far more benefits from government entitlement programs than their counterparts in the United States. They have, in effect, a government-funded social safety net for accidental injuries. The result is a system in which those injured by defective products are largely compensated by the taxpaying public at large, not by the manufacturer who made the product.[3]

In the United States, there is more reliance on the civil justice system to seek compensation for those who are injured by defective products. Generally, government and tax dollars are left out of the process, and the maker of the product pays the injured party directly.

Injured consumers clearly need such assistance. Health care costs have skyrocketed. Although private insurance and workers' compensation provide some coverage, it is often inadequate. It is not unusual for someone completely disabled, who needs around-the-clock medical assistance, to have bills of several million dollars in the first few years alone. If, for example, a 25-year-old carpenter who earns $25,000 a year is severely injured and cannot work again, the financial loss in wages alone is at least $1 million over the expected career.

COMMON MISPERCEPTIONS ABOUT
THE PRODUCT LIABILITY SYSTEM

In discussing the issue of product liability, it is necessary to get past much of the inaccurate rhetoric. First, the product liability system is not out of control. In recent years, critics have cited the figure of 18 million civil lawsuits filed every year as evidence that the product liability system has gotten out of hand. However, the great majority of those cases were small claims, divorces, probate matters, and contract disputes.[4]

Tort cases, excluding small claims, are about 2 percent of the total state court caseload and 10 percent of the civil docket. Product liability cases are tort cases, but torts also include any case in which someone's person or property is damaged and covers everything from auto accidents to wrongful death.[5] Federal court records and estimates by the Conference of Chief Justices of the states show that product liability cases have declined in recent years.[6]

Rather than product liability, other kinds of lawsuits have overwhelmed our legal systems. Corporate commercial cases—businesses suing each other—have accounted for the greatest growth in federal court lawsuits. The most serious problem with our courts, however, has come with the deluge of criminal, mostly drug-related, cases that squeeze out civil cases. Between 1990 and 1992, 10 states had to close their courts to civil cases temporarily because of the huge surge in criminal cases.

Second, product liability damage awards have not been soaring. This misconception seems to be based on anecdotes and small samples and ignores the far more thoughtful studies. The most comprehensive study so far about product liability cases was released by the General Accounting Office (GAO) in 1989. It studied all cases that went to trial in five states. Of the 305 cases studied, it found that plaintiffs won less than half the time. Even when they did win, the GAO found, many of the awards were reduced on appeal. Moreover, only 21 of the 305 cases resulted in compensatory damages of $1 million or more, and all were because the victim was either killed or permanently disabled.[7]

Recently, the Roscoe Pound Foundation studied punitive damages in state courts from 1965 to 1990. The foundation discovered that during that 25-year period there were only 355 product liability cases nationwide in which punitive damages were awarded. In fact, the odds of a U.S. manufacturer being assessed a single punitive damages award were less than one in 1,000.[8]

Third, Americans are not overly litigious in the product liability area. A recent RAND study found that only 10 percent of the victims injured by accidental injuries ever use the tort system. The report concluded that "[m]ost Americans who are injured in accidents do not turn to the liability system for compensation. . . . In this respect, Americans' behavior does not accord with the more extreme characterizations of litigiousness that have been put forward by some."[9]

Unless they are wealthy, injured persons must find a lawyer to work for contingency fees and to shoulder expenses that can be staggering in lengthy proceedings. In one study, the median cost for preparing a case for punitive damages was $30,000, excluding lawyers' time. Because of these costs, lawyers cannot afford to take on questionable cases or even many legitimate ones.[10]

Fourth, America's product liability system has not stifled innovation, led to higher prices, and caused American manufacturers to become less competitive. A system that is reported to add less than 1 percent[11] to the retail price of products cannot be blamed for America's competitiveness problems of the 1970s and 1980s. Americans do not buy Japanese cars, Korean videocassette players, and German machine tools because the U.S. liability system imposes such a heavy burden on manufacturers; the causes of U.S. competitiveness problems go much deeper. Also, foreign manufacturers do not enjoy an unfair advantage when selling their products in the United States—they must meet the same product liability standards as American firms.

The 1980s were a decade of major studies of American competitiveness. As compiled by the Council on Competitiveness, there were 14 such major studies by councils or commissions from 1982 to 1992. They included the Business Roundtable, the MIT Commission on Industrial Productivity, the Carnegie Commission on Science, Technology, and Government, the Congressional Office of Technology Assessment, the Department of Commerce, Department of Education, and four by presidential committees or commissions. In only one of those reports, by the Business Roundtable, was the product liability system cited. Even there, it was relegated to a few paragraphs in a lengthy study. In the other 13 studies, the most authoritative over the span of a decade, not a single one of them listed the product liability system as even a minor part of our nation's competitiveness problem. The most frequently cited problems affecting our competitiveness were short-term management strategies, cost of capital, technology transfer problems, worker skills, and trade barriers elsewhere.

Unlike the heavy hand of government regulation, the liability system enforces voluntary safety standards without prescribing every precise operating and technical detail. It dictates results, instead of methods, and thus encourages innovation and self-regulation. This was documented by a 1987 Conference Board report, *Product Liability: The Corporate Response*, which surveyed the risk managers of 232 large U.S. corporations. The study found that "the pressures of product liability have hardly affected larger economic issues, such as revenues, market share, or employee retention."[12] The Conference Board study continued, "Where product liability has had a notable impact . . . has been in the quality of the products themselves. Managers say products have become safer, manufacturing procedures have been improved, and labels and use instructions have become more explicit."

Furthermore, it is most telling that many of the United States' competitors are moving toward a system that is more similar to the American system of product liability. For example, the European Community adopted a directive on product liability that required member nations to enact legislation the result of which would make their systems closer to the American system.[13]

OPPORTUNITIES FOR IMPROVEMENT

Despite these benefits, the product liability system can and should be improved. The U.S. justice system mirrors U.S. society. As science and technology have advanced, new products have come cascading into the marketplace, consumer protection groups have grown, and the legislatures and courts have faced new and complex challenges. In some situations, the system has not been up to these demands. The American Bar Association (ABA) has conducted studies to look at this issue and developed the following recommendations:

1. The ABA supports narrowly drawn federal legislation for occupational diseases, such as asbestosis, when the disease has long latency periods, the damages threaten significant numbers of manufacturers, and the claims have excessively burdened the judicial system.

2. At the state level, the ABA supports many improvements in the tort system. For example:

- Annual studies of tort awards, and guidelines to encourage uniform awards.
- A tough "clear and convincing" evidence standard for punitive damages, levying them only when there is a conscious or deliberate disregard for safety.
- Strong court sanctions against frivolous lawsuits.
- Courts disallowing excessive lawyers' fees.
- Limits on joint and several liability to economic loss in certain cases. Defendants should not have to pay someone else's share of noneconomic loss they had little or nothing to do with.
- Strong court controls on excessive pain and suffering awards.

While great attention is given to many of the details of the product liability system, the greatest threat to manufacturers and consumers is the excessive costs and delays in our civil justice system. This is due in part to the lengthy process of stop-start discovery and motions. The situation is aggravated by the increase in criminal cases, which is bleeding resources from our civil justice system. The system has become too slow, too costly, and too inaccessible for most Americans.

We must have a revolution in our justice system. As a start, the ABA has met with more than 30 national organizations interested in the civil justice system. The issues discussed included ways to

1. Streamline discovery and reduce its cost.
2. Force parties to confront serious settlement discussions at the start of a lawsuit. Ninety to 95 percent of civil cases ultimately settle, so it is beneficial to settle them before substantial costs and client time are incurred.
3. Institute new procedures designed to produce earlier and cheaper resolution of matters.

As an example of the third area of change, the ABA helped pioneer a program in the 1970s called the Multi-Door Courthouse, which encourages alternatives such as arbitration, mediation, and conciliation. Today, there are more than 1,200 court-related programs that help solve entire categories of civil suits with such alternative dispute resolution methods. So far, these techniques have had limited use in product liability cases with the exception of automobile and toxic tort cases. But the potential for shifting from courts and litigation to quicker and cheaper forms of resolving disputes is great. To paraphrase psychologist Abraham Maslow, "if the only tool you have is a hammer, then every problem looks like a nail." The gentle nudge of conciliation can often spare people from the sledgehammer blows of a trial.

People from all walks of life must work together if the U.S. justice system is to be maintained and improved, and if its goal—justice for all—is to be realized. Scholars can study the problems with our justice system, lawyers can advocate change, and judges can render verdicts. But in the end, it is the public that must decide what is best. The ABA plans to take the best ideas now being developed to improve the system to the Congress and out to the states where it will encourage the creation of state justice commissions. These commissions will bring leading members of the state bench, bar, court administration, and public together to design, among other things, a faster and less expensive means of conducting the business of the courts. From their efforts is likely to come a package of improvements reflecting the concerns of the local legal culture as well as the type of improvements the ABA has discussed with various other national organizations.

Members of the legal profession share engineers' commitment to building the best products in the world and to having America compete and win in the global marketplace. America's product liability system is a competitive advantage, not a disadvantage, because it results in safer products with minimal government interference. More important, it provides a fair, open system in which consumers with legitimate claims can be protected while also shielding manufacturers against unwarranted claims.

In the long run, America will continue to succeed as an economic power because it is also a just nation in which people have the opportunity to work and succeed, and in which people can be confident that the justice system will treat them fairly. The workings of the U.S. justice system are not perfect, but its goals are. The great jurist Benjamin Cardoza stated, "Justice is not to be taken by storm, she is wooed by slow advances."[14] As the legal community "engineers" these advances, it will work with technologists and others to accomplish this important task.

NOTES

1. "R. Buckminster Fuller, Futurist Inventor, Dies at 87," New York Times, July 3, 1983, page 17. "R. Buckminster Fuller Dead, Futurist Built Geodesic Dome," New York Times, July 2, 1983, page 1.

2. Michael Rustad, Roscoe Pound Foundation, *Demystifying Punitive Damages in Product Liability Cases: a Survey of a Quarter Century of Trial Verdicts*, 23 (ed. by Lee Hays Romano) (1991) (citing U.S. Product Safety Commission, *Who We Are, What We Do* [1988]).

3. A survey by Werner Pfennigstorf and Donald Gifford comparing the liability system in the United States with that of 10 other countries suggested that the substantive liability rules in the United States do not make it easier in general for accident victims in the United States to recover for negligence. The other nations mentioned in the same study rely more on government entitlement programs than on their tort system. Although other countries may appear to have quantitatively less litigation and lower overall damage awards than the United States, accident victims receive compensatory benefits through nationalized health care, replacement income programs, and expansive workers' compensation systems, thus making resort to the litigation process less necessary.

Werner Pfennigstorf and Donald G. Gifford, *A Comparative Study of Liability Law and Compensation Schemes in Ten Countries and the United States* 158 (Donald G. Gifford and William M. Richman, eds., Commissioned by the Insurance Research Council) 155 and 158 (1991).

4. Deborah R. Hensler, *Taking Aim at the American Legal System: The Council on Competitiveness's Agenda for Legal Reform,* 5 Judicature 244, 245 (1992).

5. National Center for State Courts, *State Court Caseload Statistics: Annual Report 1991.*

6. Administrative Office of the U.S. Courts, Annual Report of the Director (1991) and Administrative Office of the U.S. Courts, Annual Report of the Director (1992).

7. General Accounting Office, Pub. No. HRD-89-99, *Product Liability: Verdicts and Case Resolution in Five States* 31 (1989).

8. Michael Rustad, Roscoe Pound Foundation, *Demystifying Punitive Damages in Products Liability Cases: a Survey of a Quarter Century of Trial Verdicts,* vi (ed. by Lee Hays Romano) (1991).

9. See Deborah R. Hensler et al., RAND Institute for Civil Justice, *Compensation for Accidental Injuries in the United States* 110 (1991).

10. Rustad, Supra, note 8 at 5.

11. Peter Reuter, *The Economic Consequences of Expanded Corporate Liability: An Exploratory Study*, v (1988).

12. The Conference Board, Report No. 893, *Product Liability: The Corporate Response*, 2 (1987).

13. On July 25, 1985, the Council of European Communities adopted the Directive. It contemplated that all necessary implementing legislation or regulations would be enacted by the member states within three years. For a discussion of action by various members states on the Directive, see Marianne Corr, "Problems with the EC Approach to Harmonization of Product Liability Law," 22 *Case W. Res. J. Int'l L.* 235 (1990). A 1988 article on the Directive notes that it "incorporates many features of product liability law in the United States." The article discusses the dominant trends in the United States as related to the EEC Directive. See Lawrence C. Mann and Peter R. Rodrigues, "The European Directive on Products Liability: The Promise of Progress?", 18 *Ga. J. Int'l & Comp. L.* 391, 393 (1988). See also Heinz I. Dielmann, "The European Economic Community's Council Directive on Product Liability," 20 *Int'l Law.* 1391 (1986). An article in the *Cornell International Law Journal* compares concepts in the Directive with European and American laws. See Marshall S. Shapo, "Comparing Products Liability: Concepts in European and American Law", 26 *Cornell Int'l L.J.* 279 (1993). For further comparison and a copy of the Directive, see Otto baron van Wassenaer van Catwijck, "Products Liability in Europe," 34 *American Journal of Comparative Law* 789 (1986).

14. Quotable Lawyer, p. 158, edited by David Shrager and Elizabeth Frost, Facts on File (1986).

IMPACT ON ENGINEERING PRACTICE, INNOVATION, AND CORPORATE STRATEGIES

The Chemical Industry: Risk Management in Today's Product Liability Environment

ALEXANDER MacLACHLAN

The ability to innovate is the key to business success in virtually every industry, but nowhere is this more true than in the chemical industry. This paper will describe how the current product liability environment has affected innovation at one company in that industry—DuPont.

DuPont participates in more than 30 major businesses based on its chemical processing skills. These businesses include agrichemicals, polymers, synthetic fibers, fine and commodity organic and inorganic chemicals, photographic materials, energy, electronic fabrication materials, and many others. DuPont is in virtually all these businesses either because the technology was invented at DuPont or because its innovative processes and products give it an advantage over its global competitors.

DuPont also has developed broadly diversified businesses outside the chemical industry that manufacture finished products such as STREN® fishing line and mammography film, medical products such as the ACA® Discrete Clinical Analyzer + Reagents, and pharmaceuticals such as Coumadin® and Percodan®. Innovation in the supply of raw materials such as chemicals, plastics, fibers, and electronics has a dramatic effect on the ability of businesses and industries to innovate in making and selling improved and competitive finished products.

RISK MANAGEMENT

The manufacture of most raw materials and finished products requires the application of risk management skills. The risks for raw materials operations typically include business, manufacturing, handling, and environmental disposal risks. Risks for finished product operations include finished product design analysis, consumer labeling, and product safety risks. These are only a few of the risks that businesses consider. For example, in the past three decades the phenomenon of injury litigation has become a major risk that is having a chilling effect on innovation in many American industries.

Chemical manufacturing processes are almost always run at high temperature and pressure and involve toxic chemical intermediates, or in some cases, extremely dangerous end products. Examples of these end products are concentrated hydrogen peroxide, which is used by the paper industry, and sodium cyanide, which is supplied to the mining industry. DuPont chooses to supply these materials based on its ability to manage risks and to protect its employees, customers, investors, and the environment.

Safety and risk management are a way of life for DuPont employees who work with chemicals. Despite the inherent dangers in manufacturing and handling chemicals, industrial safety statistics show that the chemical industry is one of the safest industries, and DuPont has one of the best employee safety records. In fact one of DuPont's businesses sells safety training to other companies, including the contractors that constructed the Channel Tunnel between France and the United Kingdom.

DuPont's risk management procedures are elaborate and ever-improving. Risk management pays, not because it prevents lawsuits, but because risk management is good business. The company benefits from providing a safe workplace because its employees are at work applying their skills instead of absent on sick leave. By shipping and handling materials and by running plants carefully, DuPont is welcomed throughout the United States and the rest of the world as a responsible manufacturer. Providing technical assistance to customers when requested helps those customers achieve success and also enables DuPont to become the preferred supplier for those customers. By helping clients dispose of waste, or even entering into partnerships with them and their customers to recycle products after their useful life, DuPont further enhances its reputation as a desired supplier. DuPont's reputation is critical to its success.

One risk management process that helps accomplish these objectives is called product stewardship. Every material that is proposed for development and eventually sold is initially and periodically thereafter subjected to a rigorous analysis for hazards related to safe transport and recommendations for safe handling and disposal. In addition, if a product, sodium cyanide, for example, has unique hazards that require a special level of customer sophistication, risk management procedures are designed to ensure that the customer meets these requirements. In these special instances, sales may be refused to customers who do not meet the risk management requirements. The frequency with which DuPont goes through this process depends on the risk category of the material.

Every business sector in DuPont has a product steward coordinator for administering the product steward program. The business manager of the product line is personally responsible for product stewardship, and every single material has a designated product steward.

The principal responsibility of the product steward is to analyze the general hazards of the basic material related to safe transport, safe handling and disposal, employee safety, toxicity hazards, and environmental impact both during development and on an ongoing basis. The steward also certifies compliance with regulations and advises customers of the general hazardous properties inherent in the raw material; reviews material safety data sheets, labels, and product literature; visits the customer when necessary; and identifies anticipated future requirements and trends and helps adjust business strategies accordingly. Product stewards are expected to be current on all aspects of these responsibilities. All reviews are formal and are applied globally. Some would contend that these practices are one positive offshoot of product liability. To the contrary, companies do these things because they are good business practices and are critical to remaining competitive.

BROAD IMPACTS OF PRODUCT LIABILITY LITIGATION

Unfortunately, injury litigation has become a routine occurrence in America. This phenomenon has been fueled by the ability of some segments of the legal profession to make hundreds of millions of dollars in contingency fees, too often by misconstruing facts in order to orchestrate popular opinion. The hidden cost of this speculation in the "legal stock market" is a chilling effect on the freedom and ability of Americans to innovate.

By definition, innovation is controversial until it is accepted as the norm. Then it becomes a success. All innovation begins with a willingness to risk failure, which is essential to innovation. All of civilization's great technological advances began with failure.

When litigation excessively punishes the risk taking intrinsic to innovation, it deters innovation. There is a far greater risk to society from making avoidance of risk at any cost the law of the land. These risks include technology stagnation, loss of competitiveness, and loss of economic standing in the world. The results may be diffuse and difficult to quantify, but the effect on each of us and on future generations will be real.

What is billed as legal compensation for a wrong done, in other words, damage allegedly done by DuPont materials or by finished products made by other companies from DuPont raw materials, results in extremely costly litigation. Whether or not the litigation is successful, there are still huge losses in time and financial resources, which chill innovation and investment. In 1993 the company's long-range R&D budget was reduced by an additional $12 million because corporate legal costs had increased by 25 percent over the year before. In addition, there is a tremendous drain on the emotional resources and the time of management and other employees who are deposed. Employees who have gone through depositions say they will never apply their intellectual and technical skills again in areas where they may have to endure the humiliation and accusations associated with these cases, which to them appear baseless and driven by financial motives.

It is often noted that the liability environment is very different outside the United States. DuPont's experience certainly supports this contention. Although 50 percent of DuPont's sales are generated outside the United States, less than 0.5 percent of the liability cases originate there.

The justification for litigation is couched in the noble-sounding objective of "deterrence of harmful behavior." It is difficult to demonstrate scientifically whether or not this purpose is being served. One unintended consequence, however, is that the chemical industry is beginning to deny new and existing materials and products to society. This in turn is affecting U.S. competitiveness.

IMPACTS ON NEW MATERIALS AND MARKETS

The following examples demonstrate how the ability to innovate into new materials and markets is being affected.

In 1989 a business opportunity, using one of DuPont's elastomer products as earthquake shock absorbers for buildings, was identified. The size of the business for DuPont was modest—a relatively small number of pounds per year at $2 per pound of material. However, due to the high likelihood that litigation would follow if an earthquake actually occurred, DuPont decided not to pursue that opportunity. The risk of litigation, not a technology problem, dictated that decision. It has become commonplace for litigation to follow any accident, regardless of the cause. It seems that someone always must be blamed or found at fault; thus, litigation becomes an inevitable risk.

Another example of the chilling effect of inevitable litigation on innovation can be found in the field of medical devices. This does not refer to breast implant litigation, in which DuPont has no involvement. Rather this refers to the current refusal of several raw material suppliers to sell any raw materials to companies or researchers working in the field of permanent medical implants such as artificial hearts, pacemakers, hip replacements, vascular grafts, and thousands of other useful finished products.

Typically, medical devices are small and lightweight, requiring only a few cents worth of raw material. For example, a single device might contain five cents worth of raw material. One thousand devices would contain only fifty dollars worth of raw material. Yet today a raw material supplier can be forced to defend itself in court for having sold a perfectly good raw material to a medical device manufacturer when the device ends up in litigation. Even if the raw material supplier is found innocent and without liability in all of the cases, it can cost millions of dollars in defense costs. The litigation tarnishes the reputation of the company, distracts employees from normal operations, and guarantees that employees will never pursue developing or selling products used in the medical device field.

This predicament has prompted DuPont to examine its policies toward selling materials used in medical applications. In 1993 the company established a new policy that prohibits sales to companies and researchers using its materials in permanently implantable devices of any kind, *unless we are involved in the design of the article and control its application.* Any other course of action would be detrimental to the future of the company. This policy will be implemented over a period that will give time for customers to substitute other suppliers or make the substitute products themselves.

As a result, any new business ideas or new concepts for synthetic material for internal-use medical devices will have a very small likelihood of getting R&D dollars. Since DuPont cannot limit or justify the risk, either through regulatory standards or reasonable assessment of potential financial liability, it will not work in these areas and will forbid use of DuPont materials or expertise to be applied in these areas. This is extremely discouraging and frustrating for scientists in a corporation with the technical capability to make all kinds of new materials.

The impact is even broader. Should DuPont strive to make substitute materials for metals in automobiles, airplanes, liquid and gas pipelines, or any use where there is some future risk of failure, especially where the failure may occur in the long term or only when millions of units are in operation under a myriad of different conditions? Obviously it is necessary to continue to innovate and accept reasonable risk. However, when even the best attempts to control litigation costs fail, DuPont will have to invest elsewhere. Sadly many of these potential investments will be made by competitors in other countries and will reach the United States only long after the improved human welfare and newly created jobs are enjoyed elsewhere.

DuPont's Kevlar® superstrength fiber business provides an opportunity for conjecture on product liability trends. Kevlar® fiber is used in all kinds of reinforcement applications, including as a substitute for asbestos in brakes and as a fiber in bullet-resistant vests. Bullet-resistant vests are a good business with modest importance as a revenue and earnings generator. However, the business is more than that to all those policemen whose lives the vests have saved—it is a matter of life and death. Since DuPont makes only the fiber and customers make the vests, it is uncertain whether the company would go into such a business today.

As mentioned earlier, product safety is important because it is vital to business success. This author has never thought that the ever-increasing demands for product safety dampened innovation. Quite the contrary— such demands are invigorating and responding effectively to them can pay enormous dividends. However, for the first time in a long career, this author is spending time looking at markets from the point of view of avoiding risks that are unquantifiable or that put DuPont in high opportunity areas that, while valuable for mankind, have too much downside if we make a mistake.

SOLUTIONS

What can be done so that the public is protected from careless activities of industry, but at the same time industry is not driven to avoid important, but initially risky innovations?

It is common business practice to assess, quantify, and gain control of business opportunity risk. The U.S. tort liability system, with its trend toward strict and even absolute liability, frustrates that practice. There seems to be a growing perception that anyone can and should sue for everything, regardless of the facts. The solutions, as I see them, must involve all members of society:

• There is a need to make the legal system more compensation-focused than deterrent-focused. When mistakes are made and no laws are violated, there should be no incentive to invent evil intent.
• The monetary incentive to achieve enormous financial gain through punitive damages and the destruction of the reputation of responsible companies and their employees must be stopped. Ridiculously high awards in the name of deterrence enrich the legal profession, but do not improve the behavior of companies.
• Standards must be set to help manage risk both before the fact and while the product is in use. Businesses should not be punished when they have met state-of-the-art standards, even if current practices are different.
• It is necessary to understand how the current system is affecting our society, particularly in terms of lost benefits from innovations not pursued.

CONCLUSION

DuPont makes outstanding materials that do remarkable things—from artificial limbs through fire-resistant materials to bullet-resistant vests.

Customers at the leading edge of innovation seek to use our remarkable materials in their products. Some succeed with their ideas, some do not. We do not claim expertise in all our customers' fields, but we work hard to make sure we represent our materials accurately and sell them to reputable customers. We look for assurance that our customers have secured the required approvals for manufacturing their products and meet appropriate standards. But, today, we find that these safeguards are no longer adequate. The consequences strike at the foundation of America's quality of life and standard of living.

Quality materials are part of the solution of many of society's problems. They are key ingredients in achieving each of the following societal goals:

- Food and shelter for an exploding population.
- Energy-efficient forms of transportation.
- Cost-efficient health care.
- Faster communications.
- A faster, more flexible defense.
- Cleaner industries for a healthier environment.
- Revitalized infrastructure to support our move into the twenty-first century.

Actions that result in a cut-off in the supply of those materials means innovation will slow. These solutions will come more slowly or perhaps not at all. Countries with different tort law systems will benefit sooner from these solutions and may indeed own them and export them to countries like the United States.

Responsible behavior of companies is driven by the simple fact that such behavior is vital for continued business success. Although irresponsible companies should be held accountable for defective products, frivolous lawsuits do not make already responsible companies more responsible. Rather they cause those companies to pull back from areas of R&D that are perceived as too risky in the current climate. In the end, the public suffers because innovations are not pursued that may make products safer and that may create better jobs and an improved quality of life.

Medical Devices, Component Materials, and Product Liability

PAUL CITRON

R oughly 40 years ago, the era of implantable medical devices was ushered in. With it, exciting new therapeutic options became available. These devices significantly complemented the medical armamentarium that was at that time limited to pharmaceutical preparations, surgical intervention largely based on excision of diseased tissue and expendable organs, and perhaps most successful of all, "tincture of time."

In the 1950s, medical devices such as large-diameter vascular grafts for the first time permitted surgeons to replace body parts that had become defective. The year 1958 saw the implantation of the first electronic device, the cardiac pacemaker. This revolutionary technology stimulated the heart experiencing bradycardia, a too-slow rate, to a rate approximating resting normal. In this way the heart was once again able to pump sufficient quantities of blood to meet the majority of the patients' hemodynamic requirements.

The 1960s saw the emergence of the mechanical heart valve as a practicable replacement for defective natural valves. Implanted medical devices such as these offered therapy where previously none was available. In many instances they offered the possibility of not only saving patients' lives but also restoring their quality of life. Perhaps the clearest example of the lifesaving capacity of medical devices has been the development of implanted defibrillators. This emerging technology is capable of detecting fibrillation as well as the dangerous heart rhythms that can lead to fibrillation, then automatically delivering a precise shock to the heart to restore normal rhythm. Carefully selected patients who receive such devices would most likely have become victims of what is aptly known as sudden cardiac death syndrome. Other technologies such as drug delivery systems, orthopedic joint implants, and intraocular lens implants have restored patients to fuller and more productive lives.

PERFORMANCE CHARACTERISTICS

It is important to note the hostile nature of the environment in which medical devices must function. Over its evolutionary cycle, the body has created a formidable set of defenses against foreign materials. It recognizes them as being potentially dangerous and vigorously sets out to consume, destroy, or isolate intruders. Obviously these defenses work well. Unfortunately, they do not recognize medical devices as allies and seek to destroy them. Only a limited number of materials such as silicone rubber, certain polyurethanes, a small number of other polymers, and an equally small number of inert metals and exotic alloys have been found to be clinically acceptable for implantation. These materials are used to construct the implant itself or serve as the protective barrier to shield other components that are not clinically acceptable for implantation.

As if the requirement of clinically acceptable tissue response was not enough, implanted materials must also be "biodurable." That is, they must have the intrinsic physical and mechanical properties to withstand the rigors of the application in which they are used. Consider two examples. A pacemaker lead that connects the stimulator to the heart must be able to maintain its integrity for years while being flexed by the beating heart approximately 38 million times per year (Figure 1). It must also withstand the abrasive action of blood and the wiping action of the heart valve through which most leads pass while still serving its primary role as a stable conduit for electrical signals passing to and from the heart. As another illustration, a prosthetic mechanical heart valve is expected to function flawlessly for the patient's lifetime while being subjected to wear forces and large pressure-induced forces as it opens and closes (Figure 2). It is not unrealistic for a valve to experience in excess of 600 million opening and closing cycles. It must do this without mechanical failure and without appreciably damaging blood cells or causing other serious complications.

These are challenging requirements that are being met with today's technologies. Even though performance is excellent, there is room for improvement. The perfect implant has yet to be developed, and this is primarily related to the effects of the blood coagulation system.

REGULATING MEDICAL DEVICES

The critical nature of medical devices has caused them to come under stringent regulation in many parts of the world. Clearance to market de-

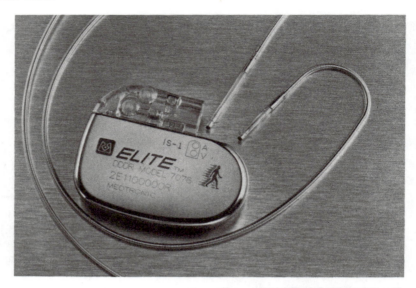

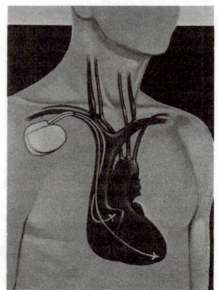

FIGURE 1   **Top:** A pacemaker system consists of two major parts—the pulse generator, and the lead/electrode which connects the pacemaker to the heart. **Bottom:** The pulse generator, which contains the circuitry and power source to monitor heart activity and produces stimulation pulses when required, is typically implanted subcutaneously in the pectoral region near the collar bone. Stimulation pulses move along the lead, through the electrode at its far end to cause the heart to contract. In a similar manner, signals produced by the heart travel through the lead/electrode to appropriately alter the pacemaker's operation. SOURCE: Medtronic, Inc.

FIGURE 2   A mechanical heart valve must perform perfectly for a lifetime while withstanding wear and pressure-induced forces. A fabric sewing ring allows the valve to be sutured into place. SOURCE: Medtronic, Inc.

vices in the United States is granted only after the Food and Drug Administration (FDA) has determined through its classification and review procedure that there is reasonable assurance of the safety and effectiveness of the device. Such regulatory requirements are necessary and appropriate. They impart a degree of discipline and thoroughness to the process. They also provide a third-party appraisal of the suitability of a new technology in comparison with other available alternatives. A rigorous but responsive and responsible regulatory process helps to ensure that new medical technologies represent the state of the art, have the real potential to do good as demonstrated in scientifically grounded studies, and reach patients promptly.

IMPORTANCE OF MEDICAL DEVICES

The impact of medical devices has been profound and far reaching. A survey conducted by the National Center for Health Statistics of the Centers for Disease Control and by the FDA's Center for Devices and Radiological Health estimated 11 million Americans in 1988 were alive with one or more implantable products, such as artificial joints, fixation devices, intraocular lenses, pacemakers, or heart valves (Biomedical Market Newsletter, 1991). This industry has global significance and is one of the few in which the United States has a positive trade balance. It is estimated that in 1992 the U.S. medical device industry produced a $4.0 billion favorable trade balance on $39.7 billion in annual sales (Health Industry Manufacturers Association, 1994). The best measure of the impact of medical device technology, though, is in how it improves and sustains patients' lives. This must never be forgotten.

IMPACT OF PRODUCT LIABILITY

Despite the enormous contribution medical devices have made to the public health, it is a business that frightens many. This fear is largely a consequence of the possibility of liability exposure in the event of device malfunction or failure. As is the case in other businesses, the specter of product liability in medicine is peculiar to the United States. Its influence is growing and is having a chilling effect on innovation. It also damages global competitiveness and increases health care costs directly and indirectly. Ironically, the shadow of product liability may actually be keeping better performing products from the market rather than being a force for improvement.

The Suppliers' Dilemma

Typically, medical device manufacturers must rely on components and raw materials manufactured by other firms in order to produce their final products. While some implants, for instance those used in reconstructive plastic surgery, consist of a single, homogeneous material, the majority of devices integrate a broad range of technologies. In many instances these components are produced for other commercial applications but have also been qualified for use in medical applications.

The dilemma that exists for the raw material or component supplier can be illustrated by the following example. Consider for a moment the mechanical heart valve (Figure 2). To secure such valves to the heart, a fabric sewing ring encircles the valve. The valve is literally sewn into place by the surgeon, who sutures through the fabric into heart tissue. Certain polyester and polytetrafluoroethylene (PTFE) fibers have proven suitable for this use in more than 20 years of clinical experience. To the best of knowledge, there have been no adverse clinical events associated with the fabric or fiber.

Yet, a key manufacturer of the fibers notified the heart valve industry in 1993 that it will discourage future use of its material in permanently implantable products. The company had concluded that selling raw material could not be justified in light of the business risk of litigation from merely having their raw material in permanent implants.

Suppliers Become Liable, Too

Precedent has demonstrated that upstream raw material suppliers can and do get pulled into product liability cases even where they had no direct involvement in the design, specification, or manufacture of the final product. They are viewed as having deep pockets and become defendants. Despite the fact that suppliers of raw materials usually win in court and are found not liable, the cost of proving themselves innocent and the management attention that must be given the matter far exceed the business opportunity. For instance, a supplier can be subjected to hundreds of lawsuits for a medical product in which its material was used as an ingredient. The material can be procured on the open market through commercial supply-house channels without any direct involvement by the supplier in the specification, design, testing, or manufacture of the end product. The out-of-pocket legal costs can, and have, run into the millions of dollars on only dollars worth of raw materials sales.

The medical device industry has seen a growing list of highly reputable material supply companies such as Dow Chemical, Dow-Corning, and DuPont announce their intention to restrict sales to implant manufacturers. These companies have sharply reassessed the extent and manner in which they participate in medical devices. This has shifted the balance in the relationship between suppliers and manufacturers. Formerly, materials were supplied by large, sophisticated chemical companies, with well-established quality procedures, to smaller companies. Litigation has moved the market to a new relationship where small, often undercapitalized start-ups with no manufacturing history provide material to the device companies. Liability concerns have driven mature, technologically well-established firms from this market.

While the impact has been greatest for implanted polymeric and elastomeric materials, it has not been restricted to them. The adverse experience with product liability has caused suppliers of essentially all components used in implants to assess their willingness to supply. For example, certain well-established manufacturers of integrated circuits have refused to supply their chips for implanted devices.

Impacts of Short-Term Solutions

The device industry is engaged in an all-out effort to find and qualify equivalent replacement materials and sources of supply. Although this process is likely to be successful, the resources expended on these initiatives will merely enable the medical device industry to pick up where it was before this occurred. The state of the art will not have been advanced. Some would suggest that the departure of the leading specialty chemical companies from selling materials to implant manufacturers may even lower the overall quality standard of the portfolio of implantable materials as smaller, undercapitalized, and less sophisticated supply sources step in and attempt to fill the void. What is clear is that the flow of new materials that would permit as yet unmet clinical needs to be addressed will be markedly slowed. The exiting companies have the laboratories from which future breakthroughs would have been likely to come. This is not to say progress has ceased, but rather, it has been slowed—and for the wrong reasons.

In many instances suppliers and medical device companies can contractually shift the risk of product failure to the device manufacturer. This is not a complete answer, however. Suppliers still can be joined in the lawsuit and must put up with the expense of discovery procedures and the great inconvenience it entails, as well as adverse publicity. While indemnification can make material available in some cases, it adds significant cost to the final product without adding any value to the physician or patient. If the device company is not financially strong, suppliers will not be comforted by such a shift in risk. This places start-up companies and entrepreneurs at a disadvantage. In this way innovation is negatively affected as are the patients who could have benefited. All of the initiatives targeted at securing continuing supply of components divert R&D dollars from activities that could provide better products to patients.

PRESCRIPTION FOR PROGRESS

The chilling effect of product liability on U.S. medical device innovation and, ultimately, competitiveness, has been outlined above. An obvious remedy is sweeping tort reform. However political reality suggests this will not occur, at least not in the foreseeable future. Perhaps, then, the following steps are a prescription for progress that will remove barriers to medical device innovation and availability.

1. The industry must communicate clearly to patients, physicians, and other sectors of the public the intrinsic limitations of medical technology. Expectations must be in line with the industry's abilities and the state of the art. Yes, medical devices are able to do miraculous things, but the perfect implant has not yet been achieved. Until it is, results will sometimes be imperfect.

2. The medical device industry has an obligation to produce high quality products, track their performance, conduct research to expand understanding of underlying mechanisms of action, and invest in initiatives that build on knowledge gained to produce evolutions of improved products.

3. The FDA, as part of its product approval process, should maintain

master files on materials that have been found to be clinically acceptable for implantation and suitable for defined applications. Medical products employing such materials and which secure marketing approval from the FDA should be deemed safe and effective as well as representing an appropriate standard of technology. Product liability actions that may be brought against the manufacturer of the end product would, in consideration of the rigorous FDA qualification process, exclude pain, suffering, and punitive damages as long as the product was produced in compliance with the terms of FDA approval.

4. Component and raw material suppliers should be shielded by law from medical device product liability actions for FDA-approved products if readily available "off-the-shelf" materials meeting specifications are incorporated into implantable products that undergo FDA approval. The burden for demonstrating suitability would fall to the manufacturer of the final product. In instances where modified or custom materials or components are provided, the supplier of raw materials would have similar protection as long as the design specification was met.

5. In those instances when individual patients cannot recover from the manufacturer costs due to device malfunction, a government-administered fund modeled after the one established for children experiencing severe complications from vaccinations would be created. In this way, the good of the many would be preserved while keeping whole those who experience an unexpected problem.

CONCLUSION

Medical device breakthroughs over the last 40 years have had a profound positive impact for millions of patients around the world. These triumphs were made possible by a spirit of discovery and the uniquely American impatience with the status quo. We traditionally want to make things better. But our litigious nature has reached such a level that it is extinguishing the spark of innovation. Methods must be implemented that remove barriers to progress so the process of continual improvement can lead to better products and better outcomes. In this way the root causes behind product liability will be reduced as well.

REFERENCES

Biomedical Market Newsletter. 1991. First medical device implant survey released. September, p. 8.

Health Industry Manufacturers Association. 1994. The Global Medical Device Marketplace Update: Markets for Medical Technology Products. Report #94-1. Washington, D.C.: Health Industry Manufacturers Association.

General Aviation Engineering in a Product Liability Environment

BRUCE E. PETERMAN

General aviation aircraft, which range from single-engine propeller aircraft to jets that fly higher than, and as fast as, commercial airliners, are a vital part of the national transportation system of the United States and of most foreign countries. The industry meets a need for business and personal transportation that cannot be filled by commercial airliners, trucks, or automobiles. Currently there are 200,000 general aviation airplanes in service meeting these transportation needs, contributing to the manufacturing and service industries in the United States, and fulfilling many training and utility roles.

The largest segment of general aviation, piston-engine-powered aircraft, is now either out of production or produced in very small numbers. In the late 1970s, that segment gained 10,000 to 15,000 new airplanes per year and accounted for more than 100,000 jobs. Now, barely 500 new piston-engine-powered airplanes are produced each year. Also, since at least 30 percent of U.S.-produced general aviation aircraft had been exported, another offshoot of the decline in production is that this important contributor to the U.S. balance of trade has been lost. The need for new aircraft is critical, not only because of the age of the fleet, since few replacement aircraft have been built since 1985, but also because enabling technological innovation could produce safer new aircraft.

A major contributor to the virtual termination of production and the most significant deterrent to rebuilding this industry is the high cost of product liability. This outcome is certainly different from the intended goals of product liability, namely, compensating for injury and encouraging safety improvements.

AIRCRAFT ENGINEERING

Among fields of engineering, aircraft engineering is unique in many respects. Since the ability to fly defied man for centuries, the fairly recent capability to fly and even extend flight beyond what was ever thought possible, makes aircraft engineering particularly fascinating. The aircraft engineering task is complex, involving multiple disciplines and principles. Furthermore, because a large number of components are furnished by outside suppliers, aircraft engineering functions are widely dispersed, making detailed technical coordination between companies essential.

General aviation aircraft engineering is also unique from a safety point of view. Safety always has been an important consideration throughout the industry. Many engineers are also pilots or passengers with their families in the aircraft they design. The result is an attention to detail even beyond that normally attributed to an engineer. Potential and actual failures have always been studied to help develop improvements in safety. In addition, aircraft design, manufacture, quality assurance, maintenance, and operation are all regulated by the U.S. government, and safety is paramount in all the regulations. These regulations are detailed not in a few pages, but in a series of books. Aircraft that are manufactured for export must also meet strict foreign requirements. Thus, safety is both an inherent aircraft engineering concern and a requirement for certification.

ENGINEERING CHANGES BROUGHT ABOUT BY PRODUCT LIABILITY

Given this attentiveness to safety issues, it would appear that product liability, and the advent of strict liability in particular, should have had little impact on aircraft engineering. Nothing could be further from the truth.

On the positive side, while it is true that aircraft engineering concern for flight safety and accident prevention has always been paramount, more attention is now focused on failure modes and effects analyses and on crashworthiness. Also, those engineers who have experienced the extreme scrutiny of depositions and trial testimony have become better engineers. Unfortunately, these benefits are offset by the negative impacts of the current product liability environment, particularly in seven areas: engineering resource allocation, documentation, joint research efforts, design, product-related publications, certification, and regulation.

Engineering Resource Allocation

Considerable manpower is being diverted from innovative and advanced design activities to the preparation of records needed to meet product liability-related demands. This includes producing support documents for various forms of discovery and preparing defense information, often for frivolous lawsuits. In some cases, such documentation consumes 20 percent of engineering staff time. Total product liability-related expenditures (outside litigation expenses and losses) can surpass 50 percent of the entire product line engineering expenditures for design, development, and certification of new or derivative products and product improvements.[1] There is an obvious, negative impact on innovation and product development.

Requests for documents and information from the plaintiff's legal counsel are usually structured in the broadest possible terms, requiring lengthy literature searches of all documents, correspondence, notes, and reports. For example, if a piston engine connecting rod failed due to a lack of lubrication, a typical request for information would most likely include any and all correspondence, certification reports, test information, service reports, and service literature relating to any and all connecting rods and oil and lubrication systems. The initial search would serve as a stepping stone from which to request additional broader information and to launch conjectural failure scenarios aimed at some alleged design shortcoming. The emphasis would not be to determine the actual cause of the failure, but to develop a chain of occurrences that might have happened and to incriminate a defendant with deep pockets. As a minimum, this causes a one-time diversion of a significant amount of engineering time in order to respond to hypothetical considerations and, at worst, could cause inappropriate regulatory changes and lengthier diversions of attention from engineering matters.

Documentation

Heightened scrutiny of the engineering process resulting from product liability cases has substantially affected the documentation of that process. Correspondence, reports, change notices, and service literature are now proofread with an eye toward downstream interpretation and implications. Older documents are particularly problematic. When those documents were prepared, they contained much of the thought processes and suppositions that went into the decisions made. No thought was given to the possibility that later they would serve as an entrance into the technical depths of the organization through cross-examination by a plaintiff's counsel. The statements in the older documents are not incorrect, just inappropriate given the latitude for interpretation allowed in the present liability environment. Thus, engineers are diverted from engineering activities to explain notes written years ago in a different environment, or to review and redraft today's notes to minimize the chance that they will be misconstrued in the future. Furthermore, effort must also be expended to provide closure to documented issues that if left unanswered could suggest a failure to respond to knowledge of a design problem. This could later be used by a creative plaintiff's expert as support for punitive damages against a manufacturer.

Another effect of legal scrutiny of old notes is the adoption and strict enforcement in some companies of a policy to destroy records not required by law or with no demonstrated company benefit. This process not only takes time, but also requires the regeneration of data when an innovative idea that has been discarded warrants reconsideration. The additional cost or time to regenerate the information may well cause that innovation or product improvement to remain undeveloped.

Joint Research Efforts

Joint industry and university research has even suffered due to the present product liability environment, and shows the far-reaching effects of this phenomenon. On occasion, university employees or ex-employees have become expert witnesses for the plaintiff against a manufacturer using the knowledge and data obtained in the joint research. Subsequently, the manufacturer may be hesitant to pursue new joint research projects, knowing that the addition of an outside party to the research team may be detrimental in a product liability context.

Design

Because of the current product liability environment, engineers are sometimes required to go to excessive lengths to "Murphy Proof" designs.[2] This reflects the assumption that engineers can anticipate and design for every possible misuse to which their products might be put. For example, one company reported that ailerons (roll control surfaces on the wings) were removed from an aircraft for maintenance and were reversed when they were reinstalled. Despite the fact that physically they were not reversible and that hinge brackets had to be deformed in order to reverse the installation, it was alleged that the design was not sufficiently "Murphy Proof."

Altering designs in response to regulatory recommendations can also be questionable. At one time the U.S. Federal Aviation Administration (FAA) published a recommended standard for locating the gear and flap controls that the pilot actuates. The location of the gear and flap controls on many of one manufacturer's aircraft that were produced prior to the recommendation did not conform. At the next change, the company revised the control locations to conform to the FAA recommendation only to have a product liability issue because the new location "confused the pilot."

Product-related Publications

The language used and amount of information given in publications such as manuals and service bulletins have been affected by the product liability environment. The "failure to warn" doctrine has resulted in a proliferation of warnings for both expected and conceivable uses or hazards. This not only consumes excessive amounts of valuable engineering time in crafting the content and language of each warning, but also generates so many warnings that they lose their impact. Also, the concern for being held accountable for encouraging possible misuse of the airplane has resulted in deleting some worthwhile information from at least one manufacturer's manuals.

Two examples of this are procedures for getting through clouds for VFR (visual flight rules)-rated pilots—those not rated for instrument flight— and soft field takeoff procedures. Instrument-rated pilots have been taught and tested to be able to control an airplane's flight path without visual reference to the ground or the horizon. VFR pilots are not licensed to fly in weather conditions that require instruments for maintaining attitude control. Some manufacturers' manuals have provided information on how to control the airplane in clouds as an aid to VFR pilots should they inadvertently be caught in clouds. Unfortunately, some lawyers have contended that such information could cause a VFR pilot to fly under conditions that, in some cases, could result in a crash. The same is argued for description of soft field landing and takeoff procedures. Thus, information that would be helpful to many pilots has been removed from the manuals.

Certification

The current product liability environment has added cost to the aircraft certification process. FAA engineers have been, and continue to be, influenced by this environment. In many cases, the certification criteria for showing compliance are now excessively conservative. This is especially true when a well-developed, rational database does not exist, and "what if" studies are required to an excess. Although undocumented, experience shows that it has become difficult to get certification approvals from FAA engineers who have been personally involved in product liability lawsuits in which a prior approval was questioned.

Regulation

New regulations are being developed for questionable safety issues, those for which there is no absolute proof that a problem would ever exist, even though aircraft have been operating safely for years without allocating engineering time to the subject or complicating the design as is now required. Examples include satisfying the new regulations for flying through lightning and high-intensity radiated fields and requiring dual load paths for mechanical controls. Not only must design engineers satisfy themselves, but tests often must be repeated for the FAA to witness. While this has always been true to some extent, the amount of repeat testing required and the need to analyze or test noncritical cases instead of relying on engineering judgment is increasing.

CONCLUSION

General aviation meets a vital need in the national transportation system of the United States and most foreign countries. However, current general aviation aircraft are becoming obsolete and are not being replaced, largely because of product liability constraints on manufacturers. The current product liability environment has caused a reluctance to include new technology in products and a diversion of financial resources from new product development. It has also added to the cost of the new product development that is being done, resulting in higher aircraft prices and reduced aircraft sales (see also Sontag, in this volume).

The general aviation aircraft engineer has been directly affected by the product liability environment in that significant amounts of time and resources that should be devoted to innovation and product improvements are being diverted to satisfy the legal requirements of product liability. As a result, a national resource—the engineering talent that should initiate manufacturing, job creation, and product export—is being wasted. The final irony is that little if any of the continuing improvement in general aviation safety can be traced to product liability litigation.

NOTES

1. These data came from a comparison of outside litigation expenses and losses with the engineering budget for a specific company's (name omitted) engine product line. Over an 11-year period, from 1982 through 1992, outside litigation expenses were at an average level equal to 51 percent of the engineering expenditures noted. On an individual year basis, this ratio varied from 23 percent to 116 percent.

2. Murphy's law: If it is possible for something to go wrong, it will.

Indirect Effects of Product Liability on a Corporation

FREDERICK B. SONTAG

Unison Industries manufactures ignition systems and other electrical components used primarily on aircraft engines, with applications ranging from Piper Cubs to 747s. It sells $50 million worth of products a year and employs 400 people in two manufacturing locations in the United States. Customers include well-known engine manufacturers such as General Electric and Pratt & Whitney, and almost every airline and corporate aircraft operator worldwide.

Despite the range of Unison's customer base, nowhere is the company's product liability exposure greater than in general aviation. Unison's product liability insurance expense for general aviation is eight times greater per dollar of sales than for all other aviation markets the company services. Of the more than 35 major product liability claims against Unison during the last 12 years, all involved general aviation aircraft. The company has yet to sustain a single product liability claim for a commercial airline or military aviation application.

Unison's experience in general aviation provides an insight into how companies change when they try to cope with product liability. The purpose of this paper is to show the indirect effects of the current product liability environment, as experienced by a small company that supplies parts to an industry with high product liability exposure.

BACKGROUND

During the 1980s, few U.S. industries fared worse than general aviation. Since 1979, light aircraft production rates have dropped more than 90 percent, and industry employment has fallen more than 50 percent. Many companies that had been active in general aviation have decided not to invest any further in that market. Industry experts blame this situation on product liability.

Piper Aircraft's experience shows how the high costs of product liability can affect a company. In 1978 Piper produced more than 5,000 airplanes and employed 8,000 people in three manufacturing facilities. Since that time, Piper has gone through several changes in ownership and massive downsizing, finally declaring bankruptcy in July 1991. The cost for defending itself against product liability claims had escalated so dramatically that by 1987 Piper was paying a premium of $30 million for an insurance policy with a deductible of $25 million. At that time, Piper had only $75 million in sales, so it was paying almost 50 percent of its revenues for product liability insurance. The expense was too much to bear.

In 1993 Piper was still operating in bankruptcy under Chapter 11, yet despite its financial condition, it had shipped almost 200 airplanes since July 1991. Because of this demand for Piper's products, attempts have been made to rebuild the company by buying it, moving it, selling off portions of its operations, or getting the liability reduced through the judicial process. Product liability has in one way or another stymied every plan, and it is questionable whether Piper will ever be able to get back on track again.

Trying to cut losses by stopping production does not necessarily relieve the burden of product liability either. Cessna Aircraft stopped making piston-engine airplanes in 1986. Yet in 1993 Cessna still paid $20 million a year for product liability expenses and was being actively sued in approximately 200 lawsuits deriving from its pre-1986 production. The commitment of resources to battle this volume of litigation still has a major effect on Cessna's ability to maneuver in the marketplace.

PRODUCT LIABILITY IMPACTS

Product Strategies

One way in which the product liability environment can affect a company is in its product strategies. This is especially true in general aviation. Not only have design improvements been slowed, but many companies, manufacturers of components as well as final products, have left the market entirely. In some cases, the void is not being filled, leaving the customer a more restricted choice in products and services.

The growth of the kit plane industry is a specific product strategy that has resulted from product liability. Rather than sell assembled airplanes, some companies are hedging their product liability exposure by selling only plans or plans along with a parts kit to be assembled. In 1993 more than 2,000 general aviation airplanes were sold in the form of kits, more than double the production of completely assembled planes. Making the end user also the manufacturer of the plane limits product liability claims for most manufacturing and some design defects. Some designers even offer with the kit an elaborate set of instructions for establishing legally separate corporations for manufacturing, owning, and operating the airplane.

Kit planes have pushed general aviation back into becoming a cottage industry again, essentially reversing the last 150 years of progress made in production techniques for this industry. The jobs and cost advantages of manufacturing complete airplanes in a centralized location have been replaced by the efforts of individuals working in their homes. In terms of safety, which plane would you rather fly—one made at Cessna or one built in someone's garage?

Relationships with Other Firms

Product liability has affected the ability of companies to acquire, divest, form joint ventures, or license manufacture. In the disposition or reconfiguration of a company or product line with high product liability, one of the major issues is how to handle that liability. While the risk can be indemnified against, there is always an issue of the nature of the indemnification or the strength of the indemnifying party. Even a sale of assets usually does not dispose of product liability. This sometimes leads to canceled deals, as happened in March 1991 when the French firm Aerospatiale withdrew from its proposed acquisition of Piper Aircraft because of inadequate product liability indemnification.

It can also lead to transactions being structured more around coping with product liability than optimizing the future prospects of a business or product line. In the late 1980s Unison was interested in acquiring a small product line with a large product liability problem. Despite a good fit between the new line and Unison's base business, the risk of combining the new business with the company's existing products proved too great to take on. As a result, a transaction was structured whereby the product line would be run in a separate corporate entity and facility that would indemnify the selling party against product liability. The proposed transaction was far from optimum for long-term economic growth of the purchased product line or stability of employment of its workers. It was, however, the only way to cope with the product liability problem. Eventually, the deal fell apart over the issue of indemnification.

Vendor relationships are also affected. Since every company in the manufacturing chain is at risk for product liability, some vendors elect not to sell to companies making high liability products. Beech Aircraft has reported that it has had to replace more than 100 vendors who dropped them as a customer, all at a very high resubstantiation cost. Other vendors drop out of production completely and cannot be replaced, which forces manufacturers to integrate vertically, often at greater cost. Some key vendors demand to be added to their customer's product liability insurance policy, again at additional cost to the manufacturer.

Another unusual twist to vendor relationships has taken place in general aviation. Since the original equipment manufacturers (OEMs) of airframes, engines, and certain major components must bear the burden of product liability, they usually sell replacement parts at a price high enough to help defray the product liability cost. Some enterprising vendors to these OEM manufacturers, who do not have the same liability burden, have begun to sell their components directly to the end users at prices below those of the OEM producers, thus undercutting their own customers. Since aviation manufacturers cannot easily switch vendors, some OEMs are forced to handle the problem by placing contractual limitations on vendors, putting themselves potentially at risk of an antitrust claim.

Financing

Product liability restricts the financing choices of a company. Most lenders are wary of lending to organizations with substantial product liability risk, even though the company may have adequate insurance. To make matters worse, it has been suggested that lenders be held liable for product liability risks of companies in which they invest, much like certain environmental liabilities. Naturally, this causes serious concerns on the part of lenders who are considering financing companies with a high product liability exposure.

Relationship with Regulators

Product liability affects the relationship of a company with regulatory authorities. In 1992 the Federal Aviation Administration (FAA) began to simplify the process by which aircraft aftermarket part manufacturers can have their products certified for repair and service use. Most OEM system manufacturers believe that this new procedure allows aftermarket manufacturers to build parts to a lower quality standard than that to which the OEM is held. These lower quality parts eventually are incorporated into the OEM system by people performing repair and overhaul. While the new FAA procedure potentially provides more parts availability in the marketplace, it also introduces an increased product liability risk for the OEM system supplier. This has resulted in a noticeable strain in the relationship between OEM producers and the major agency that regulates them.

Internal and External Communications

Product liability has had a tremendous impact on both public and internal communications. Since all statements, whether written or oral, can be construed as warranties, marketing documents have to be scrutinized carefully to make sure the words cannot be misinterpreted to take on a meaning not intended by the manufacturer. Most companies require that each major piece of marketing literature be reviewed by a lawyer, and some companies have *every* written document intended for the public reviewed by an attorney. Service bulletins and warranty statements sometimes read like legal textbooks. This is both costly to the manufacturer and confusing to the consumer, who is faced with complicated product instructions and a plethora of ominous warnings.

Internal memos and notes are similarly affected. These are usually intended to document meetings and thoughts of employees so that they can be referred to at some future date as a memory refresher. Today, memos and notes need to be drafted carefully, keeping in mind every possible interpretation of their contents. Accountants and engineers have had their words misconstrued and used against them so many times that companies have had to change their document policies. This includes requiring that certain or all documents be reviewed for wording before they are archived, starting courses in memo writing and note taking, establishing elaborate record retention policies that limit what documents are stored and how long they should be kept, and employing special staffs just to enforce these policies. The cost of all this to businesses with high product liability exposure is substantial.

Engineering Design

Engineers are some of the most creative people in a company. They need to be open minded and free thinking as they explore the complex technical concepts that eventually lead to the development of new products. However, their ability to design is seriously impaired if they are subject to restrictions in, or unreasonable second guessing of, the way they think and work.

Impacts are first felt in the design stage. Engineers are increasingly being required to "overdesign" products. This entails contemplating every possible alternative design, and carefully documenting why these alternatives were discarded. They must be prepared to explain why a product improvement is not necessarily a correction of a previous design error, become more skilled in the economics of various designs, and envision and design for misuses of products that are far beyond the realm of reasonable use.

What happens to engineers when a product liability suit actually strikes a company? They become the main targets for plaintiff's attorneys, because a plaintiff's attorney seeking to prove that a product has a flaw in design must locate and document that flaw. Usually that means going directly after the engineers who designed the product. When the lawsuit hits, engineers are overwhelmed with document production requests and interrogatories to answer. Likewise, they are subjected to days, if not weeks, of depositions during which every decision and direction taken in the design of a product is revisited by the plaintiff's attorney looking for an angle on which to build a design defect theory.

These obligations follow an engineer as long as the product he or she designed is in commerce, which could be for decades. In 1992, one of Unison's engineering managers was subpoenaed regarding a product he had worked on 10 years ago. When he arrived at the deposition, the plaintiff's attorney set in front of him a stack of laboratory notebooks, memoranda, and various other written documents generated 10 years ago. Even though this design project was just one of perhaps 50 projects that he worked on during that calendar year, he was subjected to one and a half days of detailed grilling regarding every word of every document associated with that design. And worst of all, the product he was being questioned about was not even one of Unison's. It was a design he had done while working for a previous employer.

No wonder engineers are, in increasing numbers, avoiding companies with high exposure to product liability claims. And how much safer are high liability products when the best engineers refuse to work on them? The net effect of product liability on a company's engineering is a narrower selection of engineers, a higher cost of engineering, a slower product development cycle, and the imposition of bureaucracy in an area where creativity, quick reaction, and bold thinking are the keys for product success.

Manufacturing

In companies with high exposure to product liability, manufacturing, too, has been burdened by product liability. Manufacturing processes, particularly changes, must be more carefully documented; more records of manufacturing lot traceability must be kept; and people in production jobs, just like their engineering counterparts, must learn the art of giving depositions. This makes the manufacturing of products more complex and costly.

Personnel Policies

Product liability has had an effect on personnel policies. People recruited for designing, manufacturing, or marketing high liability products must be screened to a different standard. Companies must also use much more caution in hiring or firing people. This is particularly true of engineers because of the increasing use of a company's former engineers as expert witnesses against that company. Imagine the impact on a jury when an ex-company engineer testifies against his former employer. Regrettably, too often expert witnesses' former employment matters more than the technical strength of their testimony.

For this reason, some companies have had to create unusually strongly worded employment agreements with engineers. Other companies have had to accept weak performance from engineers they fear would seek work testifying against them. In some cases, engineers have been hired back to companies that terminated them just to prevent them from becoming expert witnesses against their former employer.

Handling Risk

The handling of product liability claims and product liability insurance has greatly complicated the operation of many companies. In general aviation, there are approximately five accidents a day. Major manufacturers track every aircraft accident and review it to see if it could generate a product liability claim. Accidents involving serious injury or death are all investigated at the accident site, not only by safety authorities with the FAA or the National Transportation Safety Board, but also by trained accident investigators sent by the major product manufacturers. These activities are coordinated with product liability insurance underwriters. The net result is that massive expense is incurred by manufacturers *just in case* a product liability claim is filed.

If a product liability lawsuit is actually filed, a company must have its resources properly organized to manage the claim. In many companies, this is done by the corporate legal department, while in others, cases may be coordinated by a risk management department. Unison's experience has shown that while the legal strategy is important to the direction of the suit, the element that can win or lose a claim is the strength of the technical arguments. There have been many cases in which the plaintiff's technical arguments have no merit.

For this reason, defense of a product liability claim at Unison is led by the company's attorney and someone with a strong, broad engineering background. After a thorough review of the facts of the case, potential causes are evaluated. Once the suit is filed, it is possible to learn the plaintiff's theory for the accident, although sometimes this takes a very long time. With the plaintiff's theory on the table, technical information to rebut the theory is gathered. A variety of mechanisms—lab analyses, accident re-creations, employment of expert witnesses, and design reviews—are employed, both to refute the allegations and to propose other theories for the accident. The downside of putting engineers more directly in control of litigation is the tremendous drain on their time and the additional expense it entails. However, using this method, Unison's product liability costs have been held to a fraction of those of other companies with similar products.

International Competitiveness

As has been described, product liability costs entail more than insurance premiums and litigation defense. Product liability affects a company's product strategy; relationships with other firms; financing; communications policies; engineering, manufacturing, and personnel policies; and organizational structure. All these extra indirect costs have to be paid somehow, usually in the form of product price increases. Moreover, the bureaucratic burden imposed by product liability reduces a company's ability to react in the marketplace. This puts American companies at a competitive disadvantage relative to their foreign counterparts.

Some argue that American law applies equally to foreign companies doing business here as it does to U.S.-based companies. However, this argument falls apart when one considers what has happened in many markets once dominated by Americans. Once again consider general aviation.

Twenty years ago, American companies such as Cessna, Beech, and Piper were preeminent in general aviation. Today those companies have seen their businesses decimated by high product liability costs. Yet despite current market conditions, foreign manufacturers are taking a closer look at the U.S. general aviation market. One reason is that while a foreign plane manufacturer will have the same liability as a U.S. manufacturer for every new plane it sells, foreign manufacturers have little or no existing product base in America. Since product liability costs are proportional to the existing base of product sold, the foreign manufacturer's product liability cost is lower than that of its American competitors. As a result, a foreign manufacturer can sell its new planes cheaper. Even if a lawsuit is brought against a company based in a foreign country, conducting depositions, obtaining documents, and collecting damages is much more difficult than it is for a U.S.-based company.

REFORMS ARE NECESSARY

This situation is not what the crafters of product liability law intended. Product liability law was created to improve product safety and compensate victims of unsafe products. It was not meant to penalize conscientious companies that provide products and services vital to the U.S. economy.

Over the past two decades, product liability law exponents can point to only a handful of cases in which products were made safer as a result of product liability litigation or the fear of it. With these few safety successes has come a flood of examples of companies and products being damaged by the system. Moreover, only a small portion of the total amount expended in a lawsuit—15 percent according to some sources—is actually getting to the victims of unsafe products. The rest is paid for attorneys and other litigation costs.

Many have come to the conclusion that the system does not work as intended. It is too expensive, too complicated, and it is jeopardizing American competitiveness. As a result, serious efforts have been made to change product liability law. In 1993 these efforts included reforms that would establish a statute of repose for general aviation aircraft. The irony of not having a statute of repose is that it unfairly penalizes the most quality-conscious manufacturers, since their products have a longer life in the market. This proposal, and others like it, contain elements of common sense that seem so absent in the current product liability system.

General aviation is one U.S. industry whose demise can be traced almost solely to the product liability burden it bears. The same insidious effects of product liability are being felt in varying degrees by manufacturers in other industries. Perhaps it is time to ask the question: What U.S. industries are we willing to sacrifice to our product liability system?

The Effects of Product Liability on Automotive Engineering Practice

FRANÇOIS J. CASTAING

Some years ago, a young couple drove a compact car into a horse that was wandering down the middle of the road. The horse's body smashed into the windshield, killing the wife. A jury ruled the vehicle wasn't "crashworthy."

In another case, a woman tried to commit suicide by locking herself into the trunk of her car. She changed her mind, but it took nine days before someone found her and let her out. The woman claimed the automaker was negligent in not providing a release inside its trunks. A jury agreed, and the verdict was upheld on appeal.

In a third case, a teenage boy, a first-year driver, took his teenage girlfriend out for a drive. He had heavily modified his 10-year-old vehicle, including jacking it up and installing much larger wheels than the chassis and suspension were originally designed to accommodate. It was a rainy Friday night. The teenage driver had been drinking, and he was also speeding. Neither he nor his girlfriend was wearing the manufacturer-installed seatbelts. The vehicle left the road, and the teenage girl was ejected from the vehicle. She suffered permanent damage to her spine and ended up in a wheelchair. The girl could not sue the boy, because he did not have any money, so instead, she went after the manufacturer.

As different as these cases may be—a collision with a horse, a woman locking herself in a trunk, and an instance of severe reckless driving—they all have four major points in common: First, the vehicles involved met all of the requisite federal safety standards at the time of their design and manufacture. Second, in none of these cases was a technical malfunction deemed the proximate cause of the accident or incident. Third, all of these

77

incidents involved circumstances that most people would agree fall far outside of the normal operating range of cars or light trucks. Fourth, and most important, all of these cases pitted very sympathetic plaintiffs who had suffered horrible personal tragedies against what were seen as large, faceless, uncaring, "deep-pocket" corporations. In each case, the jury found that the vehicles involved were somehow deficient in their design, and large cash settlements were awarded to the plaintiffs.

The purpose of this paper is not to discuss whether justice was served in the aforementioned cases. Instead, this paper argues that these cases illustrate characteristics of today's product liability environment that have unintended and deleterious effects on the automotive engineering process.

PRODUCT LIABILITY IMPACTS ON AUTOMOTIVE ENGINEERING PRACTICE

The proliferation of product litigation cases in the United States raises an important question for those who design and engineer cars and trucks. In fact, it is a question that was asked in the title of an article written by automotive business writer Paul Eisenstein (1993) in *Investor's Business Daily*. The question is this: "Will your next car be designed by the courts?" The answer is, in part, yes.

Fortunately, we have yet to come to the point where American juries take a designer's pen and a computer in hand to execute a structural design or to develop automotive electronics. (Although in the past this engineer has had to listen to many lawyers claiming to know enough about automotive engineering disciplines to "teach" a lay jury what they need to know to adjudicate a product liability case.) Unfortunately, however, we have long since crossed the line where the threat of product liability litigation influences the design of cars and trucks—including those that are driven today. Ultimately, that threat of litigation includes three elements that have a big impact on the competitiveness of the American auto industry itself.

First, the threat of product liability suits inhibits the incentive to innovate. Ironically, it inhibits most dramatically the incentive to innovate safety features. Because automotive manufacturers are frequently called on to defend past product designs in the courtroom, American engineers are understandably hesitant to explore anything but *evolutionary* product designs. Revolutionary or radical new designs are nearly out of the question because they are simply too risky.

Second, the threat of product liability also creates a huge disincentive for the honest and critical evaluation of the features on current and past vehicles. Imagine, if automotive engineers had not been critical of their past work, everyone would still be driving Model Ts! In fact, no one should be more critical of a vehicle or its components than the people who designed and engineered it. In the auto industry, new product development cycles are now just 30 to 36 months long. That means that engineers must be looking at how they will improve the next generation of vehicles as soon as they complete the development of the current generation vehicle.

This is where the threat of product liability litigation has an absolutely chilling effect on the creative process and the free flow of ideas, for any document generated in an innovation-oriented corporate culture—be it a detailed proposal, a note in a day planner, or a sketch on a napkin—might be taken out of context and become evidence in a courtroom.

A hypothetical but very realistic example will illustrate this conundrum. An automotive engineer thinks she can design an improved antilock braking system. Her idea is to design a braking system that will pulse 20 times a second instead of 5 times a second. That would mean additional and more controlled braking power for the customers. In addition, her design would be less complex and expensive to build than an existing system. What she is saying is, "I think we can design a better braking system that more Americans will be able to afford in their automobiles." She puts her ideas on paper.

Any vice president of engineering would say that proposal is a dream come true. Maybe it will work, and a better, more affordable braking system that will increase customer satisfaction can be built. This kind of innovation, and the thinking that generated it, is certainly something the company wants to promote.

However, the engineer's promising proposal could also become the company's worst nightmare if it were to be faced with a product liability suit on the generation of antilock brakes currently in production. A case could be made, using the engineer's new braking system proposal as the "smoking gun," that the company knew that its old antilock braking system was deficient. Such a case could be made even though antilock brakes are not required by any federal standards. This explains why engineers and designers might fear that their ideas or criticism of current products might be taken out of context in court.

The third element of the threat of product liability is that it can actually prevent manufacturers from implementing new or improved designs in their vehicles quickly, the backward logic being that implementing a design change quickly is often misconstrued in a courtroom as an admission of a faulty design. As a result, manufacturers may be slower to implement improvements simply because of the fear that someone might contend that they knew from the start that a vehicle or component was deficient, even though the ability to make quick product changes can be a strong competitive advantage. "Different" is all too often made out to be "defective."

WHAT IS THE IMPETUS FOR MAKING CARS SAFER?

Many contend that one benefit of high settlements in product liability suits is that they force automakers to manufacture better and safer vehicles. Instead of "necessity being the mother of invention," this is the "product litigation is the mother of invention" school of thought! Experience in the real world of automotive engineering, as opposed to a theoretical one, shows that argument to be faulty.

One of the tactics used in courts is to try to persuade the jury that engineers and manufacturers are irresponsible and that they purposely and knowingly design defective products. They are painted as part of a large, faceless, and uncaring corporation that would gladly have people get hurt if it makes a buck. What this ignores is that virtually everyone who works for an American auto company drives the cars and trucks that they help to design, engineer, market, and assemble. Furthermore, their children, parents, neighbors, friends, and relatives drive those cars, too. While these people may work for large corporations, the safety and design of their products are a very personal concern.

It should also be noted that if indeed the number of product liability suits has a direct causal link with the safety and innovativeness of automobiles, the United States should have the most innovative and safest cars on the road. As the *Investor's Business Daily* article referred to earlier points out, in 1992, while Ford faced more than 1,000 product liability cases in the United States, Ford of Europe had just one product liability suit. In Japan, product liability suits are about as rare as sumo wrestling is here in the United States. Despite this, the Europeans and Japanese are known to be extremely innovative automotive engineers. Companies like Volvo and Mercedes, for instance, have long enjoyed reputations for being at the forefront of safety technology.

How can this be? A big part of the reason is that European and Japanese engineers can more freely experiment with engineering and designing better products. They do not have the fear that the work of proposing, discussing, trying, or testing ideas could be used against them in a future lawsuit. An American engineer never feels as comfortable being as critical of a car he is supposed to replace as would his European or Japanese counterpart. As someone who was born and raised in France, and then trained as an engineer and spent the early part of my engineering career in Europe, this author can verify from experience that the product litigation phenomenon is uniquely American.

CONCLUSION

It is well understood that product liability laws have a purpose. They are supposed to compensate for injury, promote safety, and penalize gross negligence. If a corporation is irresponsible, it should be held accountable. But in the United States, the situation has gone beyond punishing gross negligence. Now punishment is meted out for many risks that simply cannot be avoided when a product is produced and sold to a public that has wide discretion in how it chooses to use that product. When no distinctions are made in assigning responsibility for risk and companies are held responsible (and penalized) for *all* risk—from those attributable to the vagaries of human nature to those truly within a company's aegis—the ability to innovate, engineer, and compete is compromised.

REFERENCE

Eisenstein, Paul A. 1993. Will your next car be designed by the courts? Investor's Business Daily, July 12.

Approaches to Product Liability Risk in the U.S. Automotive Industry

CHARLES W. BABCOCK, JR.

Make no mistake about it: the U.S. product liability system is unique[1] and, after 25 years, remains entirely unimitated by the legal system of even one other nation,[2] even though all nations have had more than 25 years to observe its operation and then adopt it for themselves. It may be, as some contend, that our justice system is being replicated around the world, and well should many of its elements, such as the Bill of Rights. But the U.S. product liability system is not being replicated anywhere. Its contingent fees, blue-sky verdicts, punitive damages, and acceptance of highly suspect expert testimony remain unique to the United States.

Consider today's newly graduated engineer. Because of the globalization of engineering education, one can expect this engineer in Germany or Japan to have mastered the same fundamental engineering, scientific, and mathematical principles as the newly graduated engineer in the United States. Each new engineer will understand the laws of thermodynamics, for example, or Newton's laws of motion, and will attest that they operate in quite the same manner at any given point on the globe.[3]

How might more experienced engineers advise these new graduates? Surely they would wish them well, since they are members of a limited, precious resource: the world's supply of trained professional engineers. No doubt they would be urged to innovate, and indeed to "push the envelope," by seeking entirely new scientific knowledge. No doubt they would urge them to be creative, to dream, and to find practical ways to apply their knowledge.

Now let us consider today's new engineer near career end, rather than at today's beginning. When today's new 22-year-old engineers are age 74—perhaps retired, perhaps still practicing, but in either event still vitally interested in the profession of engineering—it will be the fifth decade of the twenty-first century, the decade of the 2040s. Given the recent, historically explosive growth in science and engineering, some or even most of the applied engineering knowledge new engineers possess today is likely to be obsolete by the year 2045. Thus, the lifelong acquisition of new professional knowledge will be vitally important.

But what would one tell 22-year-old engineers about the legal systems in Germany, Japan, the United States, and elsewhere around the world today? How can these new engineers expect to have their professional endeavors judged by legal systems during their careers?

It is, of course, dangerous to make predictions about anything, but in spite of the danger, let us deliberately look far beyond our day and think of that year of 2045, a time over half a century from now. It will be the United States' 269th year of independence. Some older Americans living then may be able to remember the celebration of the nation's Bicentennial, in 1976. Thirteen presidential elections will have intervened. Will the principles and practicalities of the U.S. product liability system be the same in 2045 as they are today, with today's damage award and other trend lines simply running straight out to 2045?

FUNDAMENTAL PRINCIPLES OF AUTOMOTIVE PRODUCT LIABILITY LITIGATION

It was Abraham Lincoln who said, as he began the "House Divided" speech, "If we could first know where we are, and whither we are tending, we could better judge what to do, and how to do it."[4]

Let us follow Lincoln's plan by first studying where "we are, and whither we are tending" in U.S. product liability law. What is the status of this law today, in 1994?

The United States, along with a number of other countries around the world, has as its legal foundation the common law of England, a system hailed for centuries as one of genius, one that offers justice carefully developed through the reasoned decisions of learned judges in particular cases, as distinguished from law dependent entirely upon the statutory fiat of legislative bodies. U.S. product liability law is a part of this system, albeit a very new one.

That this new branch of the common law remains unique to the United States, that it is notoriously confusing to the engineering community, and that it is still unsettled in several respects are not in dispute. It is also true that over the past generation, product litigation has attracted a substantial number of the best and brightest lawyers now practicing at the U.S. bar, on both the plaintiff and defense sides.

The basic principles of the American product liability system are readily applicable to the products of the automotive industry. Persons are permitted to sue automotive manufacturers and allege that injuries, typically sustained in a collision, were the result of the defective manufacture of a vehicle; or of the defective design of a vehicle; or that, even if the vehicle was designed and manufactured flawlessly, its manufacturer failed adequately to warn of a hazard incident to its use, and this failure to warn caused the injury of which the person complains.

Since the early 1970s, in the so-called crashworthiness cases, automotive manufacturers have been subject to lawsuits based on the amount of additional injury an occupant allegedly suffered during a collision by reason of a defect, even if the manufacturer had nothing to do with causing the collision itself. This means that every automotive collision can be the subject of a product liability lawsuit.

These different kinds of permissible allegations are by no means equally controversial.

Manufacturing Cases

Manufacturing defect cases not only are easy to explain to engineers and anyone else, but the law is quite well settled. Nor is the field particularly controversial among either legal scholars or practitioners: a product that is defective because it was not manufactured to specification can render its manufacturer liable where the defect causes a person to be injured.

Warnings Cases

Warnings cases can be controversial, and they certainly present numerous difficult issues. This paper is not the place to discuss them, but a principal problem is the absence of stable, known standards against which a particular warning can be evaluated. Eminent scholars like Alan Schwartz and W. Kip Viscusi have argued recently for national standards for warnings labels, and even for a uniform national vocabulary for warnings.[5] Professor Viscusi has noted that "in practice hazard warnings simply give plaintiff another test that producers can fail."[6]

Design Cases

Cases that involve product design are the most controversial part of product liability law. In design cases, the allegation may be made, and the jury may find, that the product of which the plaintiff complains was defective in design—that engineers improperly designed not only it but, by direct implication, all other products like it. Who are those whose behavior is sought to be affected by such decisions? What is the target population for this body of law? The answer is obvious: engineers.

It is necessary that engineers fully appreciate this truth. The very name "product liability" is somewhat misleading. The matter becomes plain when one considers cases brought against physicians for alleged professional errors: "medical malpractice" cases. The very word "malpractice" is a jarring one for professionals. Its lay meaning is "misconduct or improper practice," or "unprofessional conduct."[7] But what if most physicians were in the employ of hospitals or other medical organizations, so that the malpractice cases that are brought against them personally today were instead brought against their employers, as is the case with so many engineers? Would we then hear the phrase "medical liability cases" instead of "medical malpractice cases"? It may be that we would, but the result would be the same. The truth is that product liability design cases are nothing less than claims of engineering malpractice.

One of the first difficulties today's automotive engineers experience is that the automotive industry, entirely apart from product liability litigation, is heavily regulated in the United States, as is the automotive industry in all other leading nations of the world. In the United States, Federal Motor Vehicle Safety Standards, in almost 400 pages of text, state a wide variety of detailed performance requirements that every new vehicle sold here must meet. The standards are written in generally accepted engineering terms, and they are required to be objective and performance oriented. U.S. law states specifically that these standards must be promulgated in such a way that they "meet the need for motor vehicle safety."[8]

It is important, then, to keep in mind that in U.S. product liability automotive design cases, with infrequent exception, the plaintiff's position is and necessarily must be that the vehicle in question was defective even though its designers complied with all U.S. federal safety standards. What are the implications of this for engineers?

No one would think of proposing a federal automotive safety regulation that would state no rule at all. Nor would anyone propose that a federal safety regulation be written in invisible ink, so that no engineer could read it. No one would propose that a federal safety regulation be first published only years after engineering design work had been completed. Nor would anyone propose a federal safety regulation that, though written, simply makes no sense to engineers. And no one would propose an internally inconsistent regulation. Product liability design cases are confusing for automotive engineers, and highly controversial, because they can yield, for the design engineer, any one or more of these results.

The yes-or-no, liable-or-not-liable pronouncements of juries at product liability trials state no rules for engineers, as regulations do. Even if juries do have engineering reasons for their decisions—reasons that could assist design engineers only if they were stated in the form of comprehensible, technically competent engineering design or performance rules, as regulations are—juries do not announce these reasons.

If design engineers ever do manage to learn, perhaps through informal, post-trial jury interviews, just what engineering rules a particular jury applied, they will learn of the rules only years after design work on the product has been completed. Even then, jury members are likely to state their formulated rules in terms that make no sense to engineers. Finally, any manufacturer that is sued often in product liability cases—such as any domestic member of the automotive industry—will verify that one cannot harmonize for engineers the verdicts of many juries sitting in many cases, for there is rampant inconsistency among them.

Fuel tank design cases offer a good example. In one case, the plaintiff contends that the design engineers should have designed the fuel tank on the subject vehicle to be on its left side, rather than where is was, near the rear of the vehicle. In another case involving the same model, plaintiff contends that the tank should have been on the right side. In a third case, the plaintiff argues that the tank could properly have been on either the left or the right side, but not near the rear, where it was. In fourth, fifth, and sixth cases, the jury considers each of these allegations, in the same order. The juries return verdicts for the plaintiffs in each of the first three cases, but for the defendant manufacturer in the fourth, fifth, and sixth cases, without explaining their reasons in any of them. What design rule has the U.S. legal system thus promulgated to fuel system design engineers?

This is greatly troublesome. U.S. product liability litigation, in design cases, violates perhaps the single most fundamental and ancient principle of jurisprudence, because the rules of law it would impose on the group whose conduct is to be affected, that is, on design engineers, cannot be effectively communicated in advance to that group, or to anyone else. This is the difficulty with the concept of specialized courts, whose task it would be to decide whether a given design is permissible even though all engineering design work had ended years before.

Because of this, product liability design cases affect engineers in the automotive industry during the design phase of a product far less than do federal safety regulations, for which each company has extensive compliance programs in place. From the viewpoint of the consumer and the product user, the U.S. civil justice system, unwieldy and awkward as it is, serves only those few who use it. Regulation, in sharp contrast, is pervasive: it not only informs the design engineer in advance but also serves every product user.

How could we have arrived at this present position?

HOW IT ALL BEGAN: THE WAY WE WERE IN 1966

In the mid-1960s, many Americans believed or at least were told that automotive design engineers could easily and dramatically improve highway safety but simply refused to do it or were prevented from doing so by industry executives. As law professor Gary T. Schwartz described it recently:

One feature of public thinking in the 1960s was that major American corporations—and, in particular, the Big Three automakers—were economic colossi that could easily bear whatever burdens might be imposed on them by way of regulations or liability. A second feature of public opinions was that these corporations should not be held in high respect; indeed, they should be frequently distrusted.[9]

In their major, helpful book, *The Struggle for Auto Safety* (1990), Yale Law School professor Jerry L. Mashaw and Washington attorney David L. Harfst summarize their extensive study of the history of automotive safety regulation in the United States. They point out that the mid-1960s view depended on a fundamental assumption about design engineering—that there was no *technical* barrier to dramatic increases in highway safety, but only a behavioral one:

Of course, this . . . assumed both that engineering solutions were at hand or could be feasibly developed and that government was an effective agent of innovation.[10]

The 1960s criticism of the industry, and those of automotive engineers, was truly extreme. In their 1990 book, Mashaw and Harfst specifically cite the 1966 testimony of a former administration official, who spoke of "the venality of the automobile industry,"[11] suggested that "for brute greed and moral imbecility the American automobile industry has no peer,"[12] and concluded:

Part of the task of the management of public affairs in the modern world must be to take into account the fact that large segments of life will be in the hands of men of modest endowment.[13]

What were judges—very few of them engineers—to do in the midst of such a widespread public attitude? As Professor Schwartz suggests:

[T]ort judges in the 1960s and 1970s were genuinely responding to the appeal of the concept of liability for negligence or unreasonableness. That is, those judges did want car manufacturers to make proper decisions relating to crashworthiness; they did hope that design defect rules could induce the proper design of consumer and industrial products.[14]

Time and experience have ameliorated the extremity of the typical views of the mid-1960s, and in particular the belief, even among our best and brightest leaders, that the compulsion of law could force automotive design engineers readily to end highway death and injury. Of course, there has been much progress in highway safety, including automotive design, since the mid-1960s, but it has been the result of genuine scientific innovation rather than the mere implementation of fully developed mid-1960s technology, and it has been seen worldwide, rather than only in the United States. In the meantime, today's public seems increasingly to comprehend the excesses of the mid-1960s positions.[15]

In any event, U.S. product liability law developed in this atmosphere of the mid-1960s. As Professor Schwartz notes:

[M]odern tort law can be regarded as one of those ambitious programs initiated during the Great Society and then confirmed and further institutionalized during the 1970s.[16]

IS THE U.S. AUTOMOTIVE INDUSTRY AFFECTED BY PRODUCT LIABILITY LITIGATION?

The U.S. automotive industry was affected immediately and significantly by the U.S. product liability system, from its initiation in the mid-1960s, and it continues to be. All torts scholars surely would agree.[17] General Motors may have led the way, as it sought to defend the Chevrolet Corvair design cases in the mid-1960s, but Ford, Chrysler, and other manufacturers quickly were made defendants as well in a rapidly growing number of product liability cases.

The Nature of the Risks

It is axiomatic that product liability cases can present very substantial risks.[18] One of the high risk factors is the possibility, in many cases, of an award of punitive damages—those jury awards, not infrequently ranging into seven, eight, or even nine figures, that do not compensate an injured plaintiff but rather add to the jury's compensatory award a further amount designed to punish the manufacturer defendant.

A recent decision of the United States Supreme Court, in a case called *TXO Production Corporation v. Alliance Resources*,[19] demonstrates the risk. The Court refused to find that a punitive damages award more than 526 times greater than the actual damages awarded by the jury was so grossly excessive as to violate the Due Process Clause of the Fourteenth Amendment of the Constitution.

As it had in a 1990 decision,[20] the Court in *TXO* refused to establish a lawful boundary for legally permissible punitive damages awards, except to repeat that it had said in the 1990 decision, that "a general concern of reasonableness . . . properly enters into the constitutional calculus."[21] In *TXO*, a plurality of the Court determined that

> In sum, we do not consider the dramatic disparity between the actual damages and the punitive award controlling in a case of this character . . . we are not persuaded that the award was so "grossly excessive" as to be beyond the power of the State to allow.[22]

Justice O'Connor observed that all the justices at least agreed that it is *possible* for a punitive damages award to be unlawfully large.[23]

How should an automotive manufacturer deal with this risk and the other risks that lie in product liability litigation? Over the years, automotive manufacturers generally have followed the same approaches in seeking to defend against these cases.

A generation ago, the automotive industry, along with other industries, began to learn which approaches to this new kind of litigation were not successful. For example, some manufacturers thought it possible through contractual provisions to cause other companies to assume their product liability risks. But this soon proved either impossible legally or simply unwise, because a manufacturer could too easily lose control over matters significantly affecting its reputation and future insurability.

Some manufacturers sought to turn over all cases to their product liability insurers and simply let the latter deal with them. But this tended to put the insurers in a difficult position and proved unwise for the manufacturers as well, at least for those facing many product liability cases. The risks to the company were simply too great for manufacturers not to maintain full control over them. Another reason is that there were, and remain, various insurability problems: for example, punitive damages are not or may not be insurable in many states.

Manufacturers generally have determined over the years that it is necessary for them to control the selection of trial counsel and for their corporate attorneys to work with them throughout the course of each matter. In recent years, some corporations have developed sophisticated, unprecedented, and highly successful managerial control systems for product liability litigation. Two of this author's own colleagues, senior lawyers at General Motors, have pioneered recently in new, highly effective product liability cost control and litigation management techniques and have received well-deserved national recognition for doing so.

Needless, inappropriate losses and unnecessary, overly generous settlements can prompt waves of new case filings. Therefore, no automotive manufacturer today seeks to settle all or virtually all U.S. product liability cases. This is true even of the Japanese manufacturers, in whose home country—perhaps quite wisely—virtually all legal disputes traditionally are settled.[24]

On the other hand, no automotive manufacturer today would ask trial judges in every case for a full trial on the merits. One reason is that sometimes there is no dispute—the plaintiff is correct. An obvious example would be a manufacturing defect case in which it appears that a failure by the manufacturer to comply with its own engineering specifications caused the injury of which the plaintiff complains.

The only remaining course is the existing reality. Manufacturers seek to evaluate each risk, on a case-by-case basis, settle where that course is appropriate for a variety of reasons, and seek a fair trial where the allegation seems technically incorrect. In today's environment, this requires outstanding corporate counsel case managers, and outstanding trial defense counsel in every jurisdiction.

Insiders will tell you that a recent problem of increasing concern is unfair new pretrial discovery abuses, which, if permitted by judges, can present enormous burdens and even prevent meritorious cases from ever reaching trial. Many in the practice believe that the discovery rules, designed as they are to elicit truth, can be used as a powerful weapon, especially in cases where there is no causative engineering defect, and thus no genuine liability, so that counsel's motivation is never to go to trial, but rather to make discovery so oppressive as to force the defendant to settle a case of no technical liability.[25] Manufacturers must demonstrate the greatest respect for the judicial system while continuing to resist such pressures.

Trial

The trial record in the automotive industry is very good indeed. General Motors prevails in the substantial majority of product liability cases that are resolved by jury trials, and it is this author's impression that the same is true of other automotive manufacturers. Inevitably some cases are lost, even though manufacturers are motivated to take to trial only cases of no technical liability.

A particular, chronic problem is the quality of the science that finds its way into U.S. courtrooms in product liability cases, typically introduced through the testimony of expert witnesses. Many engineers in the automotive industry experienced in product liability design litigation are troubled by the scientific and engineering propositions that are too often presented to juries today.

Numerous scholarly articles have been written about this problem. Indeed, one scholar wrote recently that "Perhaps the most troubling issue confronting courts today involves the management of scientific evidence."[26] Perhaps because science and the law are not disciplines frequently pursued by the same person, many believe courts approach science with some considerable hesitation.[27] The obvious peril is that courts, in the interest of fairness or equality among the parties, or as a result of discomfort with science itself, may permit what Peter Huber has so correctly called "junk science" to enter the courtroom.

On June 28, 1993, the U.S. Supreme Court handed down a landmark decision, *Daubert v. Merrell Dow Pharmaceuticals*,[28] in which it ruled for the first time on the proper use of scientific evidence in U.S. federal courtrooms. On balance, many believe the Court's carefully considered decision may lessen the "junk science" problem.[29]

Preventive Advice

In the United States, corporate lawyers naturally seek to explain the product liability system to engineers, and to counsel them in various techniques for complying with it. In manufacturing cases, this process is straightforward enough. Quality control, the making of products to specification every time, is demanded by the increasingly discerning consumer. The engineer must understand that U.S. product liability law reinforces this demand, since liability in a product liability case can result from a mismanufactured product.

Design cases, however, present notorious professional difficulty for the corporate lawyer. What does the lawyer tell the engineer, and how, in rendering the advice, does the lawyer seek to maintain and increase the engineer's respect for the U.S. legal system? Should the lawyer advise the engineer to design every part and assembly to the standard of maximum conceivable safety, regardless of practical considerations and consumer resistance, even if the engineer purports to know what "maximum conceivable safety" might mean in actual design practice? Should the lawyer advise the engineer to ensure that his or her designs are identical to those in competitors' current products? Should she advise the engineer simply to meet government regulations?

Several things can be suggested to every engineer:

- Ask questions.
- Innovate.
- Write accurate documents.

With regard to the last suggestion, many contested product liability design cases are based on documents obtained during discovery that contain negative predictions written by the manufacturer's own engineers during the design process, speculative predictions that prove inaccurate when the products actually enter the marketplace. The single greatest strategy to minimize legal risk in design defect cases may be this: to convince engineers of the importance of accurate report writing.

DOES U.S. PRODUCT LIABILITY LITIGATION DISCOURAGE ENGINEERING INNOVATION?

In the automotive industry there is evidence on both sides of the question whether U.S. product liability law discourages product innovation.

Evidence That the System Does Discourage Innovation

Distinguished law professors have said it may,[30] and so has a recognized leader in U.S. transportation policy, Patricia Waller, director of the Transportation Research Institute at the University of Michigan, who stated at the Second World Conference on Injury Control that "Our system of product liability discourages the adoption of new technology."[31] Industry observers have said so as well.[32] Many would take the position that U.S. product liability law at the very least can discourage the development of a new, unique design obviously unlike existing products sold by competitors.

Other evidence may be found in the variety of automotive products available overseas but not in the United States. An American tourist simply walking about the street in a major European capital, or in Japan, may be surprised to see many different kinds of motor vehicles, such as small commuter or city cars that are as long as ours are wide; three-wheeled vehicles; the kinds of small vehicles a company like Daihatsu can design and sell in Japan, with far better fuel economy than anything available to U.S. consumers today; and trucks or vans optimized for special purposes. All of these have been type-approved by national vehicle safety authorities. It seems likely that these kinds of vehicles are not sold in the United States in large part because of product liability considerations. European and Japanese motor vehicle consumers clearly have a much larger range of engineering innovation from which to choose.

Further, there may be influences from U.S. product liability law and practice that affect design engineers but are difficult to quantify. It may be that vague, imprecisely defined fears created by sensationalized media accounts of spectacular adverse verdicts in product liability design cases affect design engineers, prompting them to avoid innovation in design.

The plain irrationality and unfairness of the product liability system in design cases surely tends to disorient engineers and make them apprehensive of further irrationalities. This alone could stifle innovation and creative freedom. The evidence, however, is unclear. Whether design engineers receive a counterinnovative message from the chaos of product liability design law certainly needs to be explored.

Evidence That the System Does Not Discourage Innovation

Engineers in the U.S. automotive industry, virtually all of whom seem to condemn U.S. product liability law in design cases, typically are hard-pressed to name particular design features that would be in U.S. vehicles today but are not because of this body of law. Motor vehicles are not designed exclusively in the United States. They are also designed and sold in Europe, in Japan, and elsewhere. Companies like General Motors and Ford design vehicles not only in the United States but also in Europe, the world's largest automotive market, and sell them there and elsewhere around the world. Japanese companies design vehicles primarily in Japan and sell them around the world. But only one country in the world ever has had U.S.-style product liability litigation. That country is the United States. The easy litmus test, therefore, is to observe the extent to which motor vehicles presently sold in the United States differ from motor vehicles of similar size presently sold elsewhere in the world.

In her Atlanta presentation, Dr. Waller went on to suggest that "at least in some instances, technology developed here is made available to consumers in Japan and Europe before its availability to U.S. consumers," citing antilock braking systems.[33] Her observation may be entirely correct. However, even if product liability concerns once caused such a delay, the concerns appear to have been misplaced. Antilock braking systems now are widely available in vehicles sold in the United States, and there has been no avalanche of product liability claims involving the systems. Despite occasional exceptions like the one Dr. Waller mentioned, U.S. export versions of European and Japanese models do tend to have the same innovative features as their domestic counterparts.

Automotive engineers the world over are encouraged to innovate. Intense competition in the worldwide industry ensures such a result. While European and Japanese innovations can be cited, so can American ones, such as the heads-up display adapted from military uses to permit the driver to observe vehicle speed and other information displayed at a point seemingly just ahead of the vehicle. Most U.S. automotive engineers surely would agree that innovation is encouraged in the U.S. industry, despite the unique legal system that prevails here.

With regard to the effects on the commercialization of innovation, U.S. product liability law and practice certainly prevent the outlandish advertising claims common a century ago, but they may also have some tendency to discourage the advertising of innovations until they prove themselves over time to the satisfaction of any reasonable observer. In any event, there is no evidence that outlandish, irresponsible safety advertising claims are being made today in Europe and Japan.

THE WIDESPREAD DISSATISFACTION WITH THE U.S. PRODUCT LIABILITY SYSTEM

The Automotive Engineering Reality

A distinguished federal appeals court judge in Washington, a former law professor, recently wrote as follows:

[B]ut for tort liability, producers would have inadequate incentives to compete . . . in reducing risk . . . .[34]

The judge may be correct with respect to some industries. But, at least in the worldwide automotive industry, the proposition seems demonstrably incorrect. Since no nation on earth has U.S.-style tort liability but the United States, then if it were true that "but for tort liability, producers would have inadequate incentives to compete in reducing risk," it would follow that no automotive manufacturer in Europe or Japan today has adequate incentives to compete in reducing risk, and thus that they do not so compete, and thus that cars sold today in Europe and Japan, after an entire generation of U.S. product liability litigation, are strikingly less safe than cars of the same size sold today in the United States. But this clearly is not so.

U.S. product liability expert Michael Hoenig is quoted in a 1993 article saying as much:

A lot of major safety innovations have come from European or Japanese manufacturers, where they don't have the lawsuits and liability actions we have. If lawsuits drove safer designs, you would think we Americans would have the safest cars in the world.[35]

What, for example, would a careful observer have concluded with respect to this question after a visit to the 1993 Geneva or Frankfurt Auto Shows? Safety features and technology were major themes. Volvo, Mercedes and other manufacturers focused on safety. Volvo said its concentration on safety was "part of your lifestyle," while Mercedes claimed that it is and will continue to be *the* safety leader. Audi focused on older persons with brittle bones. Current safety features displayed included reinforced body structure, air bags, including side air bags, seat belt tensioners, web grabbers, adaptive damping, antilock braking systems, traction control, child safety seating, all-belts-to-seats,[36] convertible rollover protection, air filtration, adjustable safety belt anchors, advanced navigation and mobile communications systems, night vision, and side-rear obstacle detection.

In the absence of U.S.-style product liability litigation, how could all these safety innovations be under development in Europe? The reality is that most engineers in the world automotive industry want to design safety into products, consumers demand it in any event, and few engineers seem satisfied with the *status quo*.

I asked a Saab official about this recently. He wrote:

When it comes to our product we have tried to build up an image of producing "one of the safest cars in the world," and this message is used worldwide. Consequently we are trying to have the same level of safety on our cars wherever they are sold. This level is primarily defined by the real safety need . . . .[37]

This view seems far more likely to be the typical reality for today's automotive engineers.

Doctrinal Difficulties with Product Liability Law

Conceptual difficulties have abounded in product liability design law from the beginning. For example, Professor W. Kip Viscusi of Duke has noted that the system

emerged as a mechanism for imposing more broadly based insurance coverage on firms. Although this kind of insurance serves a valuable role within the context of manufacturing defects, for which insurance is usually feasible, within the realm of design defects the usual insurance analogies break down. The liability burden imposed by design defects is too great to be easily spread across all consumers.[38]

Though there are legal scholars who still would fiercely defend the existing U.S. system in design cases, many others now seem willing to state in public that this emperor may have no clothes. Consider, for example, this view expressed by Professor Alan Schwartz:

Only plaintiffs' lawyers like today's products liability law . . . . [Its] foundational assumptions are either false or not sustainable on the evidence.[39]

Mashaw and Harfst, too, speaking of the existing civil liability system in the United States, suggest that "virtually no one—save perhaps trial lawyers—is content with it."[40] Professor Michael Wells wrote in a recent article:

[A]fter the vast changes of the 1960s and 1970s, almost no one is happy with contemporary tort law. Liberals believe it is still too restrictive and want to abolish it in favor of an insurance scheme to compensate victims of accidents, while conservatives think current tort doctrine already favors the plaintiff too much and would cut back on liability.[41]

Two leading legal scholars have suggested that American judges, on their own, have been changing the system for the better. In fact, they have suggested in recent articles that a "revolution" occurred in U.S. product liability law during the 1980s:

We posit that a pro-defense revolution began in the early to mid-1980s and continued through at least 1989. We base this assertion on declining plaintiffs' success in products litigation, on pro-defendant trends in explicit lawmaking in products cases at both trial and appellate levels, and on steadily declining products filings in federal courts.[42]

They have found a "widespread, independent shift in judicial attitudes,"[43] and that "the declining trend in plaintiff success is matched by an equally striking but more recent decline in products filings."[44]

But, in the author's view, there is no evidence of such a "revolution" in General Motors' product liability experience, and presumably in that of other automotive manufacturers, of "declining plaintiffs' success in products litigation," and, on average, little or no decline in new case filings, whether it be on the federal or the state level, over the past decade. The seriousness of product liability cases, on average, seems to be steadily increasing. Discovery abuse is definitely on the increase as well, and this of course drives up the costs of product liability litigation.

Although these are indisputably distinguished, leading law professors, their proposition, at least for the automotive industry, does not seem to be so, and the industry is not likely to have missed it. The news of this "revolution" would have been very welcome indeed for shareholders, executives, engineers, and everyone else in the U.S. automotive industry at any time during the past decade.

What are the stated purposes of the U.S. product liability litigation system?

As one legal scholar said in a recent article:

The substantive rules of tort law exist to serve certain social purposes. The most prominent among these are compensating innocent victims for injury and deterring behavior that presents risks that exceed their social value.[45]

The first concept proposes that the legal system ensure that people injured in the course of their use of a product should be paid money. Note that there is no insistence or requirement that the injured should have their *medical* costs paid. If this were so, then the proceeds rationally would be payable directly to their doctors, hospitals, and other health care providers. Rather, as Stephen Sugarman has pointed out, the insistence is that the money be paid directly to the injured:

[T]he award is normally paid out in a lump sum (a clear advantage to the lawyer), rather than in periodic payments the way that Social Security, workers' compensation, private disability insurance, and health insurance are paid.[46]

Is the U.S. system efficient in these two objectives? It is certainly not efficient in the first. Professor Sugarman has noted that "Personal injury law is staggeringly inefficient as a system of victim compensation,"[47] and, as to the second objective, that "There is little reason to assume that it importantly curtails unreasonably dangerous conduct . . . ."[48]

He concludes:

If tort law fails as a behavioral control mechanism, is it justified as a mechanism for compensating accident victims? On this score, the current system is ludicrously inefficient . . .

since the "costs of litigation, primarily lawyers' fees, roughly equal what claimants receive as compensation."[49]

Will the solution for all this come soon, either through the "tort reform" movement, or in decisions handed down by existing courts? For a discussion of these issues, see the appendix to this paper.

ROLE OF THE ENGINEERING COMMUNITY: THE MISSING CONTRIBUTOR

It is vitally important to our national competitiveness that the spirit of technical innovation be encouraged and permitted to flourish. Is it in the interest of U.S. society in the twenty-first century to deter our engineers, especially at the beginning of what promises to be an explosion in worldwide engineering innovation? Will the rest of the world be deterring its engineers? Would anyone a century ago have thought it appropriate to deter Edison?

The engineering profession does not appear to have been centrally involved in the continuing intellectual debates over the future of U.S. product liability law and practice, especially in design cases, even though, as noted previously, the allegations in these cases are of engineering malpractice. Legal scholars have been centrally involved in these debates for years, as have economists, but engineers generally have not.

Should a comprehensive study of the desirable future of U.S. product liability design litigation be constructed, it might conclude that today's U.S. product liability litigation system provides a good, fair way to judge the design work of professional engineers. But perhaps it would not. The study would certainly address the question of whether U.S. product liability, this creation of legal academia and the bench, is fair and appropriate from the viewpoint of the engineering community.

Any comprehensive study of the propriety of U.S. product liability design litigation for the automotive industry certainly would begin with an attempt to define the actual, underlying problem.

The Real Problem: Highway Safety

U.S. product liability law has been an attempt to respond to a continuing, though progressively declining, national tragedy: death and injury on our highways as the result of the use of motor vehicles. To be sure, death and injury occurred on American roads in previous centuries, but the unprecedented personal freedom made possible by self-propelled motor vehicles in this century also has brought with it unprecedented numbers of highway deaths and injuries. This has been by no means limited to the borders of the United States but has been a worldwide problem.[50] It is to this problem that the legal systems of every nation with substantial numbers of motor vehicles have sought to respond during most of the twentieth century.

The ultimate, perhaps *twenty-second* century solution is for engineering essentially to eliminate human injury in the ground transportation modes. The highway statistics tell us that this necessarily requires the engineering of solutions that will correct or avoid errant driver behavior, the overwhelming cause of highway death and injury. It may be that today's experimental Intelligent Vehicle/Highway System (IVHS) technologies[51] will someday prove to have been crude, early precursors of such a day. If this occurs, it will make manifest the mid-1960s dream, but not because of lethargy, venality, and incompetence on the part of the engineering community. Rather, it will be for quite the opposite reason: because engineering will have completely overcome human error, in ways acceptable to the citizenry, and in ways that preserve or enhance the existing freedom of personal mobility.

The principal cause of highway death and injury, although perhaps politically inconvenient, is the same around the world. All the studies have shown, and continue to show, that most highway death and injury is due to driver behavioral factors,[52] with only a small remaining portion due either to highway design or to defects in vehicles, most of the latter in used vehicles by reason of inadequate maintenance.[53] If a distinguished engineer were suddenly appointed czar of automotive safety in the United States, with vast new budgets and pervasive powers to make laws, he would not, given the statistics, rationally concentrate the vast resources entrusted to him upon defects in new vehicles, although he might spend some of his time and resources addressing defects in *used* vehicles. Surely, however, he would devote most of his resources to the overwhelming leading cause of highway death and injury, driver behavioral factors.

He would most likely begin with the single leading cause of highway death and injury, drunken driving,[54] and the second leading cause, the failure to wear the safety belts that are present and ready for use in virtually every vehicle on the road today. Of course, he would include highway design in the scope of his efforts as well. Doubtless he would be pleased to learn that highway fatality rate in the United States is not constant, but rather that it has been steadily declining.[55] Surely the near future offers the promise of continued declines.

In summary, overall highway safety is the fundamental problem the law seeks to address. But U.S. product liability law, much federal safety regulation, the safety critics, and the attention of much of the press, all for the past 25 years have focused on a tiny, statistically infinitesimal part of highway safety: defects in new vehicles.

THE ULTIMATE, TWENTY-FIRST CENTURY SOLUTION

The current product liability trend lines clearly cannot continue unabated for the next 50 years. If they were to do so, the system—one which many would characterize as a lottery even today—surely will implode, having failed from its own excesses, long before the year 2045. Profound change, therefore, seems inevitable.

If the engineering community were freed from the weight of an enormously expensive and inefficient U.S. litigation system that considers disputes only years after the design process is over, and even then yields few if any useful messages to guide design engineers in their professional conduct; if the engineering community were thus fully freed to innovate; if a successful national health care system were well-established (see the appendix); then what, ideally, would remain? The answer seems to be an enlightened, engineer-centered federal regulatory system.

As Professor Schwartz has suggested:

In theory, society could resolve the product-defect problem by regulation; in fact, resource limitations prevent the state from regulating more than a small subset of products and product warnings.[56]

But why give up so readily? Well before 2045 A.D. the United States seems likely to have adopted such a regimen, at least in major industries such as the automotive industry, where well-established regulatory schemes already are in place. Professor Gary Schwartz believes that even modern liberal legal thought reaches the same conclusion:

Leading liberals . . . seemingly favor the abolition of tort law, so that it can be replaced by expanded systems of social insurance and safety regulation.[57]

In his book, Professor Viscusi suggested:

The past two decades have witnessed the establishment of a series of regulatory agencies designed to promote product safety. In a world in which we have . . . a National Highway Traffic Safety Administration . . . it makes little sense to have juries making sweeping regulatory decisions by assessing design defects issues on the basis of the features of a particular case.[58]

He therefore proposed the following:

Firms should be exempted from potential liability in design defect cases if they can demonstrate . . . compliance with a specific governmental regulation . . . .[59]

And indeed, regulations are written; they are promulgated well in advance of the engineering design process; and they are required to be comprehensible and internally consistent. This may be why regulation, and not civil litigation, is the norm in every other nation in the world.

Does regulation stifle innovation? Since regulation governs known technologies, perhaps it is true that it does not necessarily provide an impetus to innovate. But regulations requiring specific engineering designs for consumer use obviously should not mandate unknown, unproven technology. Performance-oriented regulations, such as the safety regulations that apply to the U.S. automotive industry, in fairness should follow the same rule. Engineering innovation, however, is inevitable and will continue, quite apart from regulation, on a worldwide basis. The obvious answer is that as innovation occurs, regulations requiring change should be changed. There are well-established processes to facilitate such changes, and they seem to work reasonably well all over the world. If on occasion a U.S. automotive regulation does stifle innovation, then surely the proper remedy would be to amend the regulation appropriately rather than turn the whole subject over to an extremely expensive and wasteful civil litigation system unique to the United States.

It is this author's belief and prediction that well before today's new engineers retire, there will be a profound revolution in U.S. products law. By 2045, engineering-oriented federal regulation will deal with product engineering design as necessary for the benefit of the public, but the product liability civil litigation system as an adjunct to this will be obsolete. The regulatory process will be effective and enlightened in ways even beyond those of which today's progressive agenda thinkers dream. This means that the twenty-first century will feature the complete absence of product liability jury trials, blue-sky jury verdict potentials, punitive damages, and highly suspect expert testimony. If this seems radical, one must remember that it is the norm today, in 1994, in every other country in the world.

Just as workplace injuries were removed from the U.S. tort litigation system at the beginning of this century because they were not fairly manageable within it, product liability design cases—at least those involving the products of comprehensively regulated industries—will be removed from the civil litigation system, for the same basic reason.

Finally, when all this has occurred, and it is indeed the year 2045 A.D., suppose one were present in a leading university classroom. How is a 2045 engineering professor, or a law of torts professor, likely to describe U.S.

product liability design case law and practice as it was in the late twentieth century? Will the professor tell the students it was an efficient, rational system that was fair to those whose conduct it sought to affect—that is, to design engineers? Will professors tell students the system achieved its avowed purpose, in that, from the clear vantage point of half a century later, the system caused products sold in the United States to be dramatically safer than those sold anywhere else in the world? These and similar questions seem to answer themselves today.

AREAS FOR FURTHER STUDY

It was Francis Hutcheson who wrote "That action is best which procures the greatest happiness for the greatest numbers."[60] How can the United States best use its limited supply of trained, professional engineers to procure "the greatest happiness for the greatest numbers" of its citizens? Indeed, how can a peaceful, post–Cold War world of rapidly increasing population best use the world engineering community for the same purpose?

The importance of engineering and technology to our national welfare as the world enters the twenty-first century cannot be overestimated. Only international cooperation in engineering and technology will permit the world community fully to reap the benefits of rapidly increasing technological progress.

Therefore, there is a need to study not only current trends in the interpretation of product liability law and how organizations respond to them but also the ways in which the legal systems of all advanced nations, including, but not limited to, the United States treat the alleged malpractice of professional engineers.

What seems to be needed is a study of world legal systems *from the viewpoint of the engineer.* The engineer, more than anyone else, is in a position to describe the legal framework that would be, for the worldwide engineering profession, fair, reasonable, and just. Such a study surely would include the extent to which legal rules that permit societal judgments to be made about the professional endeavors of a design engineer can be communicated to and understood by design engineers, in advance, before the design process begins.

A well-designed, objective study on this subject surely would be of great value to the Congress and to all Americans. It is possible to conclude that today's American system is quite acceptable, or that in any event the subject is too controversial to be addressed. But it is not possible to conclude that today's U.S. product liability law and practice is and will remain unimportant to the engineering community.

APPENDIX: REFORMING PRODUCT LIABILITY LAW

Tort Reform

There has been a great deal of recent activity in both the federal and state legislatures designed to reform the U.S. product liability legal system.[61] Indeed, one law professor has written that the tort reform movement, by the mid-1980s, was

the most active period of statutory reform of tort rules in western legal history.[62]

Tort reform is legislative in concept rather than judicial, in that the law is to be reformed by the enactment of statutes rather than by the considered decisions of common law judges in particular cases. Legal scholars can detest this because the process seems political:

[T]he reasoning set forth in judicial opinions contributes to the intellectual development of tort law. Tort-reform statutes, by contrast, are typically lacking in any such effort at reasoned explanation.[63]

Although statutes certainly can set forth rules of law without much ancillary explanation, such complaints are somewhat misleading. This complaint does not concern mere tort reform proposals, or the debates over them, but rather tort reform *statutes*: bills that have been passed by legislatures despite vigorous opposition from plaintiff lawyers and others, and signed into law by governors. In virtually every such case, the legislative process generates a voluminous public record. The "reasoned explanation" for each new statute, therefore, is readily to be found in the recorded chorus of complaints over the effects of the existing tort system that created the widespread support necessary for its successful passage.

Will today's "tort reform" movement provide the ultimate, mid-twenty-first century solution? It can help, to be sure. The subject is so significant that in our company, one experienced senior product liability expert and lawyer has been put in charge of the field nearly full-time. A number of leading litigation experts at major national law firms work virtually full-time today for tort reform. The importance of significant change to help restore fairness in product liability litigation makes these efforts correct for our time.

But what of 50 years from now? The ultimate, twenty-first century solution seems much more likely to be comparatively radical, well beyond the boundaries of "tort reform" as the term is understood today. Consider, for example, this view:

Although more than forty states adopted tort reform legislation of some kind during the last decade, on the whole this legislation has merely tinkered with tort law doctrine and cannot be seen as fundamental change.[64]

Although affected manufacturers surely would protest the use of the word "tinkered" to characterize legal change that can have very significant effects in existing litigation, it is nevertheless true that tort reform is not likely to produce "fundamental change."

A typical misgiving about "tort reform" is that the judges who created the legal system that tort reform statutes are designed to repair must themselves interpret the statutes. If judges are not convinced of the need for reform, their decisions may tend to be conservative, restrictive, and not in the spirit intended by the reformers.

Court-Led Reform

But what of reform led by these judges themselves? What are U.S. judges likely to do on their own about the present state of U.S. product liability law and practice? Are they the ultimate, twenty-first century solution? In this regard, two leading law professors, Professors Henderson and Twerski, suggest that the "truly interesting question is . . . what limits courts will set on design litigation."[65]

Judges, of course, *could* be the necessary reformers, and in the future they may well be. Their judicial predecessors created the U.S. system, after all, and today's judges are free to reform it. Professors Henderson and Twerski predict that, going forward, courts

will be "leaner and meaner" than . . . in the seventies and early eighties. Courts, assisted here and there by legislatures, will shift more of the responsibilities for managing generic risks to product users and consumers.[66]

But what of the opposite direction? Will the courts "correct" the problem by imposing "enterprise liability"—that is, manufacturer liability for any injury related to the use of one of its industry's products, regardless of cause? There seems to be little interest in this blatantly unfair change in the law, one that would establish a hopelessly inefficient compensation mechanism. Professor Henderson seems correct in stating that "I agree with Schwartz's conclusion that sweeping enterprise liability is not part of the future of American tort law,"[67] and that ". . . court-made strict enterprise liability would be totally (and unfairly) unmanageable."[68]

How can any legal system better promote safety and still encourage engineering innovation? The most likely answer is that by 2045, and hopefully long before, the United States will have brought itself into line with the legal systems of other advanced nations. Ironically, this is, at least in general concept, what several leading torts scholars seem to have predicted 40 years ago.

The *Georgia Law Review* recently featured an important symposium on "Modern American Tort Law."[69] The symposium included articles by several of today's leading torts scholars, including Gary Schwartz, David Owen, Michael Wells, Kenneth Simons, and James Henderson.

Modern readers are indebted to Professor Schwartz for pointing out in his article that the January-February 1953 issue of the *Northwestern Law Review*[70] had likewise featured a symposium on "The Law of Torts." The 1953 symposium included articles from several of the leading torts scholars of that day, including Leon Green, Fleming James, Clarence Morris, Albert Ehrenzweig, and Fowler Harper. Today, in 1993, one can evaluate their ideas against the reality of the subsequent 40 years of U.S. product liability law development.

As Professor Schwartz has noted:

[Fleming] James' ultimate proposal was that tort law should be abolished in favor of a more general system of social insurance (combined with improved programs in safety regulation).

An indeed, in their classic 1956 casebook, Harper and James wrote:

The best and most efficient way to deal with accident loss . . . is to assure accident victims of substantial compensation, and to distribute the losses involved over society as a whole or some very large segment of it. Such a basis for administering losses is what we have called social insurance.[71]

and

Beginning with workmen's compensation in 1910 and getting great impetus from the depression of the 1930s, social insurance legislation has grown apace in America.[72]

Although they admitted that

social insurance certainly rejects the limitations of the fault principle and it has for that reason been condemned as "offending the sense of justice."[73]

Professors Green and Ehrenzweig, as Professor Schwartz noted in his 1992 article, "urged the repudiation of the tort system and the adoption of social insurance; indeed, Ehrenzweig dismissed the tort system as a neurotic mess."[74] Ehrenzweig's 1953 views may seem quite sound to today's engineer:

[W]hile deterrence would, indeed, presuppose a "wrongdoer's" fault at least in the eyes of those to be deterred, it cannot support a fault liability of lawful enterprise. Clearly, imposition of liability on the manufacturer for harm caused by his defective merchandise to the ultimate consumer despite all possible caution, is not designed to deter him or others from operations otherwise so effectively encouraged by society. Nor can, realistically, a higher premium he might become obligated to pay in consequence of greater losses, cause him to exercise greater case.[75]

Ehrenzweig argued that

ultimately, in accordance with schemes proposed in Scandinavia and Germany, the development must lie towards the wholesale substitution for tort liability and liability insurance, of loss insurance . . . rather than liability.[76]

His 1953 concerns about tort litigation seem strangely timely:

[I]t is the more imperative to seek a way to remedy what has become a meaning-less game in our courts, which, by encouraging skillful and often devious practices in influencing witnesses and juries, by permitting the perversion of court trials into frivolous gambles and by preventing our judges from attending more speedily and effectively to other duties, threatens further to increase dangerous disrespect for court procedures and court law.[77]

Will some of the published views of these earlier legal scholars become a reality before their centennial? If one postulates that the good of the gen-eral citizenry would be best served by (a) optimally safe product designs and (b) universal health care, at least for catastrophic injury, then one must appreciate the reality: these two needs are well understood and are being met today, entirely without U.S.-style product liability litigation, in *every other advanced, leading nation in the world except the United States.*

We, however, have no national health care system, as the dean of the Columbia Law School and another prominent legal scholar pointed out in a 1993 article:

The United States does not have a *system* for compensating the victims of illness and injury; it has a set of different institutions that provide compensation. We rely on both tort law and giant programs of public and private insurance to compen-sate the victims of illness and injury. These institutions perform related functions, but the relationships among them are far from coherent. Indeed, the institutions sometimes work at cross-purposes, compensating some victims excessively and others not at all.[78]

The United States lacks an intellectual structure to undergird its web of programs compensating the victims of illness and injury.[79]

They suggest that

it is possible that with the advent of universal health insurance, reducing the scope of tort liability would find more political favor than at present, and that the savings from this reform could be used to help finance the health insurance system.[80]

and that

Once a system containing these elements of compensation is developed, the role of the tort system in compensating the victims of illness and injury could be de-em-phasized. Because compensation for health care expenses and lost income would already be assured independently of the tort system, the desirability of providing compensation for these losses through tort recoveries would substantially de-cline.[81]

If, by the year 2045, the United States has a comprehensive, affordable, and universal health care system, how then is it likely to address the need for vehicular safety? The most likely answer seems to be by enlightened, scientifically sound regulation. But rational, successful intermodal trans-portation policy for the mid-twenty-first century seems likely to involve the engineering community as the centerpiece, not as a target for abuse, and not an afterthought.

The proper, ultimate resolution must support and encourage the engineering profession in the United States. Nearly everyone seems to agree that as a nation we must dramatically increase the emphasis on science and technology in our educational system if we are to remain a world leader in the new century and not collapse into a mere shadow of what we were. This means the encouragement of the study of mathematics and science from an early age, especially for present minorities, an increasing percentage of America and thus of the future American workforce. The ultimate resolution obviously must encourage the American engineering profession in general and American engineering innovation in particular.

In the automotive sector, then, the public interest seems to lie not in preserving and increasing the role of the civil litigation system, but rather, as Mashaw and Harfst have suggested, in

achieving the greatest economically beneficial reduction in motor vehicle deaths and serious injuries consistent with politically acceptable levels of regulation.[82]

Peter Schuck of Yale Law School assured his readers, in a recent article on the subject of legal complexity, that the existing product safety legal system in the United States is indeed relatively complex. In his view, this is because the product safety system is, among other things, "institutionally differentiated," itself a complex term:

A legal system is *institutionally differentiated* insofar as it contains a number of decision structures that draw upon different sources of legitimacy, possess different kinds of organizational intelligence, and employ different decision processes for creating, elaborating, and applying the rules. Product safety, for example, is institutionally differentiated in that it is governed by statutory provisions, regulatory standards promulgated by several different agencies and private technical organizations, tort litigation, and common law contract principles.[83]

The difficulty with institutional differentiation, in Schuck's view, is that it

spawns legal indeterminacy, another governance cost. The proliferation of policy-making institutions multiples the sources of innovation, information, and legitimacy—precious resources in any social system. On the other hand, this diversity also encourages conflict and raises decision costs.[84]

Automotive engineers typically do not complain of the U.S. product liability litigation system in design cases by employing Professor Schuck's terms—that is, on the ground that the system is "institutionally differentiated" and thus "spawns legal indeterminacy." But it is what they mean, and it is why they criticize the system.

Professor Schuck notes that

[L]ess institutional differentiation might reduce the legal indeterminacy that such differentiation tends to spawn. In tort cases, for example, technical standards issued by regulatory agencies and satisfying certain conditions could be made presumptively binding on juries.[85]

Groups that are targets of legal systems—design engineers in this case—naturally resent institutional differentiation, since it makes the law seem incomprehensible, as Professor Schuck points out:

But if the complex legal landscape contains many pitfalls for the governors, it is *terra incognita* for the governed . . . the density of the legal system—the penetration of law into every corner of human life . . . is bound to be a source of deep resentment. . . . When this Delphic law also emerges from an institutional black box that is itself dense and difficult to comprehend, its legitimacy—the sense of "oughtness" that the lawmakers hope will attach to it—is diminished.[86]

Professor Schuck argues that the current tax structure is a notorious example of the price society pays for needless complexity, but that:

there is ample evidence of delegitimation costs in fields other than tax. A RAND study of corporate responses to modern product liability law, for example, found that the law emitted such noisy, random, and confusing signals to manufacturers that it had little effect on the product design decisions it was supposed to influence.[87]

Further, he notes that

the main producers, rationalizers, and administrators of law—legislators and their staff, bureaucrats, litigants, lawyers, judges, and legal scholars—generally benefit from legal complexity while bearing few of its costs. On balance, they prefer a complex system . . . .[88]

In contrast to the cost bearers, the beneficiaries of complexity can drape themselves in lofty public interest goals, such as securing the individual's right to a day in court, preventing the shrewd from circumventing the law, and heading off problems before they arise.[89]

In summary, we live in a period of widespread dissatisfaction with the existing system, and thus in an unsettled period.

NOTES

1. See, e.g., *The Liability Maze: The Impact of Liability Law on Safety and Innovation* (Peter W. Huber and Robert E. Litan, eds., 1991), at 47-51; Alfred W. Cortese, Jr., and Kathleen L. Blaner, *The Anti-competitive Impact of U.S. Product Liability Laws: Are Foreign Manufacturers Beating Us at Our Own Game?*, 9 J.L. & Com. 167 (1990); at 179-180; Charles W. Babcock, *Could We Alone*

Have This? Comparative Legal Analysis of Product Liability Law and the Case for Modest Reform, 10 Loyola L.A. Int'l & Comp. L. J. 321 (1988), at 351-355.

   2. Cf., e.g., James A. Henderson, Jr., and Aaron D. Twerski, *Stargazing: The Future of American Products Liability Law*, 66 N.Y.U. L. Rev. 1332 (1992), at 1341 (footnotes omitted):

> We are the only country on the face of the globe that provides such high damages in tort cases. Other legal systems view our awards as outrageous and unconscionable.

   3. Sir Isaac Newton himself once said as much:
[T]o the same natural effects we much, as far as possible, assign the same causes . . . . As the descent of stones in *Europe* and in *America* . . . . I. Newton, *Rules of Reasoning in Philosophy*, in Sir Isaac Newton's Mathematical Principles of Natural Philosophy and His System of the World 398 (F. Cajori ed., 1960)(A. Motte trans. 1st ed. London 1792)(as translated from Newton's original Latin)(emphasis in original).

   4. Abraham Lincoln, Speech before the Illinois State Legislature, June 16, 1858; in *A Treasury of Great American Speeches* (Hawthorn Books, New York, 1959), at 73.

   5. W. Kip Viscusi, *Reforming Products Liability* (1991), at 248: The genesis of this proposal is the hazard warnings section of the American Law Institute report prepared by Alan Schwartz and myself.

   6. *Id.*, at 211.

   7. Webster's New Twentieth Century Dictionary 1091 (William Collins Publishers, 2d ed. 1979).

   8. 15 U.S.C. §1392(a)(1988):

> Each . . . Federal motor vehicle safety standard shall be practicable, shall meet the need for motor vehicle safety, and shall be stated in objective terms.

   9. Gary T. Schwartz, *The Beginning and the Possible End of the Rise of Modern American Tort Law*, 26 Ga. L. Rev. 601 (1992), at 615.

   10. Jerry L. Mashaw and David L. Harfst, *The Struggle for Auto Safety* (1990), at 63.

   11. Traffic Safety Part 2: Hearings on H.R. 13228 before the Committee on Interstate and Foreign Commerce of the House of Representatives, 89th Congr., 2d Sess. 1314 (1966).

   12. *Id.*

   13. *Id.*, at 1317. There was more. The highway safety problem could best be resolved merely by changing the behavior of automotive executives:

> An attraction of this approach is that it [can] be put into effect by changing the behavior of a tiny population—the forty or fifty executives who run the automobile industry. *Id.*, at 1314.

These executives, of course, included then and include now many automotive engineers. Note the key assumption believed by so many Americans at the time:

> Probably the most efficient way to minimize the overall cost of accidents is to design the interior of the vehicles so that the injuries that follow the accidents are relatively mild. *Id.*, at 1312.

   14. Schwartz, *id.* n. 9, at 640-641.

   15. Cf., e.g., the following conclusion of an editorial that appeared in the Colorado Springs Gazette Telegraph (Friday, May 7, 1993, p. 27):

> It's often said that ours is a litigious society, and nowhere is that more evident than in the absurd expectations we often heap upon manufacturers in the field of product liability . . . .

   16. Schwartz, *id.* n. 9, at 619.

   17. See e.g., *id.*, at 632:

> Car manufacturers loom large in the jurisprudence of products liability....

   18. See, e.g., Book Note, *The Judge as Shield*, 105 Harv. L. Rev. 1124 (1992), at 1124: we live in "an age of incalculable liability exposure."

19. 509 U.S. ____, 113 S.Ct. 2711, 125 L.Ed. 2d 366 (1993).

20. *Pacific Mutual Life Ins. Co. v. Haslip*, 499 U.S. 1, 111 S.Ct. 1032, 113 L.Ed. 2d 1 (1990).

21. *Haslip*, 499 U.S. at 18, quoted by the Court at 113 S.Ct. 2711 at 2713.

22. 113 S.Ct. 2711 at 2722, 1993 U.S. Lexis 4403, *34.

23. 113 S.Ct. 2711 at 2731, 1993 U.S. Lexis 4403, *63:
It is thus common ground that an award may be so excessive as to violate due process.

24. See, e.g., *Could We Alone Have This?* id. n. 1., at 333.

25. One must resist the temptation to cite specific examples, even where it would be ethically proper to do so. This sort of thing is unpopular, especially among scholars, who typically dislike "mere anecdotes," deriding them with description like "captivating little stories." Michael J. Saks, *Do We Really Know Anything About the Behavior of the Tort Litigation System —And Why Not?*, 140 U. Pa. L. Rev. 1147 (1992), at 1161. But it is the sum of these "mere anecdotes" that constitute the whole of the existing reality. It therefore may be valuable for torts scholars deliberately to gather "mere anecdotes" from existing cases regarding current discovery practices and carefully analyze them.

26. David L. Faigman, *Struggling to Stop the Flood of Unreliable Expert Testimony*, 76 Minn. L. Rev. 877 (1992), at 883.

27. See, e.g., Faigman, *id.*, at 883:
Despite its pervasiveness, courts approach expert scientific evidence inconsistently and with trepidation.

28. 509 U.S. ____, 113 S. Ct. 2786, 125 L.Ed.2d 469 (1993).

29. E.g., see generally Peter W. Huber, *Galileo's Revenge: Junk Science in the Courtroom* (Basic Books, 1991), as well as the Huber contribution in this volume.

30. See, e.g., Stephen D. Sugarman, *The Need to Reform Personal Injury Law Leaving Scientific Disputes to Scientists*, Science, May 18, 1990, at 824 (footnotes omitted):

> . . . liability concerns are said to cause . . . enterprises to withdraw socially desired products from the market. . . . Liability jitters may discourage firms from undertaking important research or from bringing to market beneficial inventions . . . .

31. Patricia Waller, *Injury Control: How Does It Fit In the Emerging National Transportation Program?*, Second World Conference on Injury Control, May 20-23, 1993, Atlanta, Georgia.

32. See, e.g., "How Liability Stifles Design," *Design News*, February 1, 1993, at 118, in which Robert L. Rickert, President of Indramat, a division of Rexroth Corporation, is quoted as saying that the U.S. product liability situation is serious enough:

> to either scare American companies away from product development or to prolong product introduction while engineers laboriously test products to death . . . .

33. Waller, *id.*, n. 31.

34. Stephen F. Williams, *Second Best: The Soft Underbelly of Deterrence Theory in Tort*, 106 Harv. L. Rev. 932 (1993), at 933.

35. Paul A. Eisenstein, "Will Your Next Car Be Designed By the Courts?", *Investor's Business Daily*, July 12, 1993, at 4.

36. "All-belts-to-seats" is automotive industry jargon for a lap-shoulder safety belt system design in which all the belts are anchored to the occupant's seat rather than to the floor or elsewhere in the vehicle, such as the roof rail.

37. Letter from Tore Helmersson, Saab Automobile AB, Trollhättan, Sweden, dated July 1, 1993, written at request of the author.

38. Viscusi, *id.*, n. 5, at 209.

39. Alan Schwartz, *The Case Against Strict Liability*, 60 Fordham L. Rev. 819 (1992), at 819.

40. Mashaw and Harfst, *id.*, n. 10, at 241.

41. Michael Wells, *Scientific Policymaking and the Torts Revolution: The Revenge of the Ordinary Observer*, 26 Ga. L. Rev. 725 (1992), at 725 (footnote omitted).

42. James A. Henderson, Jr., and Theodore Eisenberg, *Inside the Quiet Revolution in Products Liability*, 39 UCLA L. Rev. 731 (1992), at 743-44; see also *The Quiet Revolution in Products Liability: An Empirical Study of Legal Change*, 37 UCLA L. Rev. 479 (1990).

43. *Inside, id.*, at 734.

44. *Id.*, at 742 (footnote omitted). But see, for an analysis that takes issue with the statistical bases for the Henderson and Eisenberg findings, Arthur Havener, "Not Quite A Revolution In Products Liability," Manhattan Institute, Judicial Studies Program, 1992.

45. Saks, *id.* n. 25, at 1150.

46. Sugarman, *id.* n. 30, at 825.

47. *Id.*, at 823.

48. *Id.*

49. *Id.*, 825 (footnote omitted).

50. See, e.g., Leonard Evans, *Traffic Safety and the Driver*, at 2-3, 332-338 (Van Nostrand Reinhold, 1991).

51. Patricia Waller, director of the Transportation Research Institute, recently described seven of these presently new technologies: Near Object Detection System (NODS); Run-Off-The-Road Warning; Cooperative Intersection; Adaptive Cruise Control; Collision Warning/Avoidance; Night Vision Enhancement; and Call For Help: Mayday. *Id.*, n. 31.

52. See, e.g., Evans, *id.* n. 50, at 92-93, and works cited by Dr. Evans, including B. E. Sabey and G. C. Staughton, *Interacting Roles of Road Environment, Vehicle and Road User in Accidents*, Fifth International Conference of the International Association for Accident and Traffic Medicine (1975); B. E. Sabey and H. Taylor, *The Known Risks We Run: The Highway*, in Societal Risk Assessment — How Safe is Safe Enough?, at 43-63 (R. C. Schwing and W. A. Albers, eds., Plenum Press, 1980); K. Rumar, *The Role of Perceptual and Cognitive Filters in Observed Behavior*, in Human Behavior and Traffic Safety, at 151-165 (R. C. Schwing and W. A. Albers, eds., Plenum Press, 1985); O. M. J. Carsten, M. R. Right, and M. T. Southwell, *Urban Accidents: Why Do They Happen?*, AAA Foundation for Road Safety Research (1989); J. R. Treat, *A Study of Precrash Factors Involved in Traffic Accidents*, The HSRI Research Review (May-August 1980).

See also *Report in Support of Department of Transportation and Related Agencies Appropriations Bill*, 1994, H.R. Rep. No. 149, 103rd Cong., 1st Sess. 103 (June 22, 1993):

> *Numerous studies of crash causation have shown that driver-related factors account for about 90 percent of all highway crashes.*

53. *Id.*, Mashaw and Harfst note that, at most, 13 percent of accidents involve some mechanical failure and that:

> *Within the 13 percent most failures result from inadequate maintenance, not from defective design or construction.*

Mashaw and Harfst, *id.* n. 10, at 11 (citing the Tri Level Study: Institute for Research in Public Safety, *Tri Study of the Causes of Accidents*, 7-9, 18-23 [1979] (study prepared by private consulting group for NHTSA).

54. See, e.g., H.R. Rep. No. 149, *id.* n. 52, at 101:

> *Alcohol-impaired driving remains the single most important problem in highway safety and will require sustained and intensified attention.*

Mothers Against Drunk Driving (MADD), in its 1993 annual summary of statistics, notes that in 1992

> *An estimated 17,699 persons died in alcohol-related traffic crashes—an average of one every 30 minutes . . . it is estimated that 1.2 million persons were injured in alcohol-related crashes—an average of one person every 26 seconds . . . . About two in every five Americans will be involved in an alcohol-related crash at some time in their lives.*

Mothers Against Drunk Driving, A 1993 Summary of Statistics: The Impaired Driver Problem (1993), at 1.

55. See, e.g., H.R. Rep. *id.* n. 52, at 97:

In 1992, the nation experienced the lowest ever number of highway fatalities despite an increasing amount of travel on the roads. NHTSA's latest estimates show that 39,500 persons died on the nation's highways in 1992, 1,962 fewer fatalities that in 1991, and the lowest level in 30 years. The fatality rate set another all-time low in 1992 at 1.8 fatalities per million vehicle miles traveled.

56. Alan Schwartz, *id.* n. 39, at 819.

57. Gary Schwartz, *id.* n. 9, at 695.

58. Viscusi, *id.* n. 5, at 210.

59. *Id.,* at 128.

60. Francis Hutcheson, *Treatise II, Concerning Moral Good and Evil,* in An Inquiry into the Original of Our Ideas of Beauty and Virtue, Sec 3, VIII, at 177 (London, 1716, Yale University Library, reprinted 1971, Garland Publishing Company).

61. The phrase "tort reform" is somewhat imprecise. The law of torts, or civil wrongs, covers a wide range of unreasonable human activity that courts permit to be the subject of recovery in civil litigation, include unwelcome personal advances, defamation, tortious interference with business activity, and many more activities. Only one of the many branches of the law of torts is to be understood by the current phrase "tort reform": the tort of selling a defective product that has caused an injury of which the plaintiff complains. The current phrase thus would more precisely be rendered "product liability law reform."

62. Michael Wells, *id.* n. 41, at 725 (footnote omitted).

63. Gary Schwartz, *id.,* n. 9, at 682.

64. Kenneth S. Abraham and Lance M. Liebman, *Private Insurance, Social Insurance, and Tort Reform: Toward a New Vision of Compensation for Illness and Injury,* 93 Colum. L. Rev. 75 (1993), at 76, n. 2.

65. Henderson and Twerski, *Stargazing, id.* n. 2, at 1334.

66. *Id.,* at 1334-1335.

67. James A. Henderson, *Why the Recent Shift in Tort?,* 26 Ga. L. Rev. 777 (1992), at 778 (footnote omitted).

68. *Id.,* at 781 (footnote omitted).

69. 26 Ga. L. Rev. 601 *et seq.* (1992).

70. 47 NW. U. L. Rev. 751 *et seq.* (1953).

71. *Id.,* at 762-763.

72. 2 Fowler V. Harper and Fleming James, Jr., *Law of Torts* (1956) 759 (footnote omitted).

73. *Id.,* at 761, citing Oliver Wendell Holmes, Jr., The Common Law 96 (1881).

74. Gary Schwartz, *id.* n. 9, at 635.

75. Albert A. Ehrenzweig, *A Psychoanalysis of Negligence,* 47 NW. U. L. Rev. 855 (1953), at 865.

76. *Id.,* at 871 (footnotes omitted).

77. *Id.*

78. Abraham and Liebman, *id.* n. 64, at 75.

79. *Id.,* at 85.

80. *Id.,* at 99.

81. *Id.,* at 117.

82. Mashaw and Harfst, *id.* n. 10, at 205.

83. Peter H. Schuck, *Legal Complexity: Some Causes, Consequences, and Cures,* 42 Duke L. J. 1 (1992), at 4.

84. *Id.,* at 21.

85. *Id.,* at 49.

86. *Id.*, at 22-23.

87. *Id.*, at 24, citing, in n. 96, George Eads and Peter Reuter, *Designing Safer Products: Corporate Responses to Product Liability Law and Regulation* 27-29 (1985).

88. *Id.*, at 26.

89. *Id.*, at 27.

Innovation, Engineering Practice, and Product Liability in Commercial Aviation

BENJAMIN A. COSGROVE

T he design, manufacture, certification, and maintenance of commercial aircraft constitute a complex process that has as its cornerstone the safety of the passengers and crew. The aircraft must also satisfy the economic needs of the consumer and the airline. This paper will describe how the primary goal of safety is obtained, how engineering practices have changed to obtain increased safety, and how product liability affects our design and maintenance practices.

Early jet transports were limited in speed, payload, and range. The DeHavilland Comet, manufactured in the United Kingdom, became in 1953 the first commercial jetliner in service. Early versions had a crew of five and could fly 44 passengers 2,500 miles. Although this aircraft greatly advanced the capability of air travel, it also had technical problems that caused several catastrophic accidents and undermined confidence in the airplane such that it was grounded. Despite the fact that the Comet was five years ahead of the competition, the British airplane industry never recovered after the accidents. By the time the problem was corrected, other models, mainly the Boeing 707 and Douglas DC-8, had learned from the lessons of the Comet and incorporated fail-safe features into the design, thereby garnering most of the sales for that market.

Around 1958, as jet-powered air transportation became more common and began to phase out propeller-driven transports, there was a dramatic improvement in both efficiency and safety. Each generation of jet transports saw improvements in design and production techniques. Not all the improvements, however, were in the machinery. Advances in weather forecasting, navigational aids, air traffic control, crew training, and maintenance combined with improved equipment to create safer systems. Figure 1 shows the decline in hull-loss accidents since 1960.

PRODUCT DEVELOPMENT CYCLE

Product development and the incorporation of innovations in aircraft are a complex process. As new airplane programs are initiated, design teams study existing and emerging technologies that will further improve safety, reduce weight, provide operational efficiencies, and simplify production. Before any decision is made to implement these new technologies, they are assessed against Federal Aviation Administration (FAA) rules and regulations that have been developed over the past 50 years based on data from commercial and military experience in weather, structure, engines, electrical, electronic, and mechanical systems. New technologies are subjected to developmental tests. For example, if a new material is being considered, samples from many batches are exhaustively tested for static and fatigue strength, corrosion resistance, and crack propagation rate.

Trade-off studies involving factors such as safety, weight, cost, and the economies of operations are also made. One of the questions that faces designers is how much risk to take in using a new philosophy of design or advanced technologies. Failure analyses must show that the chance of a single failure or combination of failures that result in the loss of the airplane is one in one billion or less. The risks taken are based on data and

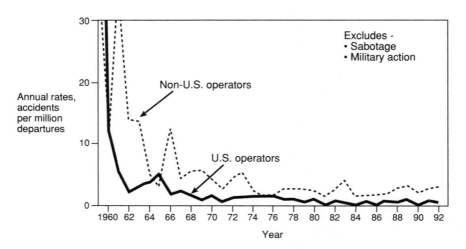

FIGURE 1   Hull-loss accidents in worldwide commercial jet fleet.
SOURCE: The Boeing Company.

knowledge that are the state of the art at the time of the risk-taking decisions. In the commercial airplane industry, federal aviation regulations also dictate that any single failure must be such that the airplane can be landed under the conditions in which it would be forced to land.

Once new ideas "earn their way" onto the airplane, that is, they are proven to provide some benefit for the passenger or the airplane, they are again tested by further component qualification tests and then by ground and flight tests leading to certification. During this phase, the manufacturer works with the airline and its crew members, particularly when entirely new models are developed, to ensure that any innovations are designed to meet their specifications. In-service shortcomings of previous models, such as failures resulting in delays and flight cancellations, high maintenance workload, and incidents and accidents,[1] are also studied. These studies provide valuable lessons on what does not work well.

The product development cycle from the program go-ahead to certification now takes between four and five years. This process takes about a year longer than it did 20 years ago, mainly because of increased regulations and more ground and flight testing. The design must meet or exceed federal regulations in effect at the time application is made for certification of a new product, although often the FAA will impose special conditions or require compliance with new regulations imposed after the time of application if significant in-service experience or new technologies warrant it. Even though the design is based on current regulations and state of the art, when an accident occurs 20 years later, the design may be criticized for not meeting present-day standards.

After the airplane is certified, it can be delivered to the airlines. The airlines also go through rigorous processes proving to the FAA that they have competently trained crews and maintenance and inspection programs in place before they receive operating approval to carry passengers on the airplane. For example, it takes an average of two months for a pilot to become qualified to fly an airplane that is a new model.

This process illustrates how the airplane transportation system is made up of three independent parties—the regulatory agencies such as the FAA, the airlines, and the manufacturers of aircraft (see Figure 2). All three parties must do their jobs properly or the system will fail. One of the FAA's roles, for example, is to provide surveillance of manufacturers and airlines to ensure that all regulations are met or exceeded throughout the lifetime of the airplane. As operators experience flight delays, cancellations, diversions due to mechanical difficulties, or maintenance difficulties, they report them to the manufacturer through its worldwide network of field service representatives. The manufacturer makes engineering changes that are developed to correct the problems and incorporates the changes into production. Service bulletins are released to facilitate changes for the air-

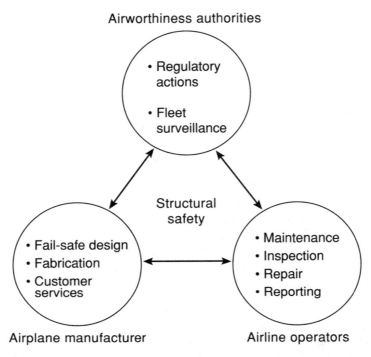

Airworthiness authorities

- Regulatory actions
- Fleet surveillance

Structural safety

- Fail-safe design
- Fabrication
- Customer services

Airplane manufacturer

- Maintenance
- Inspection
- Repair
- Reporting

Airline operators

FIGURE 2   Airplane transportation system. SOURCE: The Boeing Company.

planes in service. If flight safety is affected, the FAA can release an Airworthiness Directive (AD), which makes the change mandatory.

Figure 3 compares the causes of accidents in 1940 and today. The percentage of accidents due to engine and airplane failure has declined and the overall total accident rate has been reduced dramatically. Despite these positive trends in accident reduction, initiatives are continually being pursued to reduce accidents. In the areas of refused takeoffs (RTOs)[2] and controlled flight into terrain (CFIT),[3] for example, there is aggressive activity to reduce accidents. While the goal of zero accidents may be unattainable, it is the responsibility of the designer, the operator, and the regulatory agency to strive for that goal.

RESPONDING TO PROBLEMS

Today, with about 11,500 commercial jet airplanes in service, getting information to and from the field is a large task. Monitoring maintenance, which is the FAA's responsibility, is even tougher. A 1988 accident in-

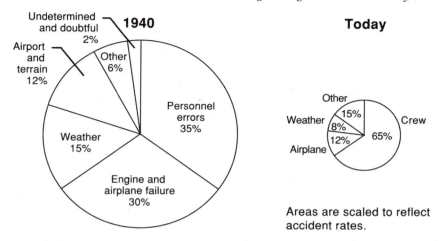

FIGURE 3   Causes of accidents in U.S. commercial aviation. SOURCE: The Boeing Company.

volving a Boeing 737 demonstrates this difficulty. The airplane was leveling off after reaching its assigned cruising altitude when an 18-foot-long portion of the upper half of the fuselage separated from the airplane. After a thorough investigation by the National Transportation Safety Board (NTSB), which is primarily responsible for investigating airplane accidents, it was determined that the probable cause was failure of the airline's maintenance program to detect the presence of significant disbonding and fatigue damage, which ultimately led to the failure of the lap joint. The airplane had been delivered in 1969 and had accumulated 35,496 flight hours and 89,680 flight cycles (landings) at the time of the accident. It was the highest cycle airplane in the fleet and was operating well beyond the anticipated service life for which the airplane was designed.

Because it was becoming increasingly common for airplanes to continue in service well beyond their expected service life in years, flight hours, and flight cycles, an industry-wide task force was formed to address the issue of aging aircraft. In this case the government, the airlines, and the manufacturers all cooperated to address the problem, but it was not the threat of litigation that caused this to happen. It was the desire and drive of the industry to maintain the continued airworthiness of the fleet and to ensure public confidence in the industry. One result of this effort was a change in the operating principle that with proper inspection, an airplane can fly indefinitely. Current guidelines encourage periodic replacement of parts instead of relying solely on extensive inspections.

IMPACT OF PRODUCT LIABILITY

How has the industry changed in response to product liability trends? The compelling reason for improvements and innovations in the aircraft industry is to maintain the reputation and public trust of the industry, not to allay product liability fears. This is not to say, however, that companies do not spend substantial amounts of resources defending themselves in litigation arising out of accidents. Much of this expenditure unfortunately does nothing to improve safety.

Although litigation often arises after an accident or incident, engineers are urged not to let that affect the work that needs to be done. A 1975 letter, now periodically reissued, from one airplane manufacturer encourages engineers to communicate improvements, safety considerations, problems, design changes, and changes in the state of the art, and not to let the prospect of litigation, or the concerns of in-house legal staff, stifle the exchange of ideas. It reads, in part, as follows:

Despite this situation, we must preserve the free flow of information within the company. That is, we should take care not to let the prospect of litigation prevent us from communicating with one another—in writing where necessary—about improvements, safety considerations, problems, design changes, and changes in the state of the art.

A far greater concern for the aircraft industry is the aftermath of media coverage when an accident occurs. Although the average number of deaths per year (approximately 130) that occur in the United States in commercial aviation are far fewer than those associated with bicycles (approximately 1,000) and motor vehicles (approximately 40,000), the nature of commercial airline accidents and the resulting media coverage cause a much different perception by the public. Typically following a major airplane accident, the news media will give front-page coverage to the accident for five to seven days followed by coverage on the back pages for another week. Based on the last 10 years of experience and today's accident rate, an airplane accident occurs every 24 days. By the year 2010, with expected fleet growth and the present accident rate, there could be an accident every 10 days. Since the media can carry news of an accident from 5 to 14 days, there will be almost continuous reporting of accidents.

Although media coverage of accidents is expected, sensational and misleading stories can create pressure on the industry that is counterproductive to public safety. This is particularly evident when the publicity and political interest that result from media coverage of an accident precipitate calls for extensive inspections that are superfluous and even foolish. On such occasions, the operators may need to open up systems in airplanes, find nothing amiss, and err in restoring the airplane to its original state, thus creating a hazard. Although the industry is strongly committed to necessary inspections, it recognizes the dangers from those that are not.

CONCLUSION

Innovation in design both in current and future generations of airplanes is part of the engineering culture. Accidents and exposure to product liability are minimized in the following ways:

• Designing redundancies into the airplane structures so that if a structural element fractures, a backup load path will carry the loads. Maintenance inspection requirements are provided that will detect the structural problem within a reasonable time so that structural integrity is maintained.
• Designing redundancies into airplane systems so that if one system fails, a backup system can operate essential functions. For example, aircraft have multiple hydraulic, electrical, pressurization, and navigational systems.
• Making changes as service experience dictates. When a significant problem occurs in service, it is tracked. It may be a single event, but if a trend develops and other operators start having the same problem, then studies can be initiated to study alternative solutions. Thus, data are compiled and tracked so that trends are monitored and management is made aware of how the airplane is operating in service.

The aircraft industry moves large numbers of people and amounts of freight efficiently and safely. It has come a long way but must continue to improve the safety and economics of the airplane.

NOTES

1. An "incident" refers to an equipment malfunction or navigational error. An "accident" refers to hull loss, and generally, loss of life.
2. Refused takeoffs (RTOs) refer to takeoffs that should have been aborted safely below V-1 speed and are not, resulting in accidents. During this period, "go-no-go" decisions can be made safely. However, above V-1, pilots are committed to takeoff, and if they do not, there may be an accident.
3. Controlled flights into terrain (CFITs) are accidents resulting from navigational and altitude errors. Many of these accidents occur on airlines from countries where the regulatory agencies do not require ground proximity warning systems on the planes. Because the instruments are deemed unnecessary, the airlines request the manufacturer to remove them.

Regulation, Litigation, and Innovation in the Pharmaceutical Industry: An Equation for Safety

MARVIN E. JAFFE, M.D

In 1992 Andrew Wiles of Princeton University demonstrated that there is no answer to the centuries-old theorem that French mathematician Pierre de Fermat left dangling in the margin of a notebook: $x^3 + y^3 = z^3$. Indeed, it represents an "impossible triangle." Perhaps now he has time to direct his analytical powers toward another scientific dilemma: Does the threat of future liability restrict innovation in the pharmaceutical industry?

Unraveling the impact of the tort liability system on innovation in America is a difficult task for any industry, whether it is aviation, toys, or soap. But pharmaceuticals present an especially complex challenge. Like Fermat's equation, the question itself is unanswerable unless reference is made to the impact of the federal regulatory process that governs every aspect of the prescription drug cycle and makes reduction of liability risk intrinsic to new drug development. The effects of the tort liability system on innovation in pharmaceuticals cannot be calculated without factoring the pervasive role of the U.S. Food and Drug Administration (FDA) into the equation.

What is the relationship, therefore, between regulation, liability exposure, and pharmaceutical innovation? Are they three sides of an impossible triangle? To the contrary, the regulatory process helps create an environment where exposure to liability is far less a factor in pharmaceutical innovation than human and medical benefit are. Consider the fact that research is ongoing into the development of new products in challenging categories such as birth control, or that thalidomide, perhaps the most infamous drug in history, is currently being "rehabilitated" for use against such devastating diseases as leprosy and AIDS.

CONSTRAINTS DESPITE SAFETY NET

Despite the regulatory safety net, however, the threat of litigation imposes constraints. Persistent, sometimes frivolous litigation casts a shadow on certain critical medical categories, notably vaccines. Even when no adverse judgments are made, companies bear the heavy costs of litigation and must pay high insurance premiums—or, most likely, self-insure. Overdesigned to reduce risk, the regulatory process allows little opportunity for serendipitous discovery once compounds enter the pipeline. Regulation also exacts a heavy toll on innovation through high costs and a plodding pace.

The asymmetrical system that arises from the divergent goals of drug regulation and the law of product liability is another key industry consideration. This paper explores this asymmetrical system and how it sets pharmaceuticals apart.

MEDICINE AND POISON

The most singular feature of the pharmaceutical industry is its products. Pharmaceuticals are designed for an intended medicinal effect, but as complex products, they may have unintended effects as well. In fact, the term "pharmaceutical" derives from the ancient Greek *pharmakon*, which literally means both "medicine" and "poison." Pharmaceuticals are categorized by the legal community as belonging to the class of "unavoidably unsafe products" including vaccines, blood, and medical devices, which offer desired benefits but are not without risk.[1] The law recognizes that the medical value of pharmaceutical products differentiates them from other products such as lawnmowers or household cleaning products. The law characterizes drugs as "not unreasonably dangerous because they are incapable of being made safe for their intended purpose."

INDUSTRY SPECIFIC SAFEGUARDS

While the law recognizes pharmaceuticals as "not unreasonably dangerous," the regulatory process is designed to protect consumers from excessive risk. No other industry in the United States has such extensive government oversight of the testing, formulation, manufacture, marketing, and distribution of their products. No other country has an equivalent regulatory system in terms of the extent of its control.

Drug regulation in the United States is pursued through several different avenues. New drugs must receive FDA approval before they can be marketed in the United States. The clinical research process is subject to direct FDA monitoring. Product labeling and promotional materials require regulatory approval as a prerequisite for sales.

Another distinguishing feature of the industry is the indirect way prescription drugs are sold. Pharmaceuticals are prescribed and dispensed by physicians and health care providers, or "learned intermediaries." Manufacturers bear the responsibility to educate the professional community about the risks and proper uses of products. In fact, manufacturers have the continuous legal obligation to "utilize methods of warning which will be reasonably effective"[2] and are "required to keep abreast of the current state of knowledge of its products as gained through research, adverse reaction reports, scientific literature, and other available knowledge" (Fern and Sichel, 1985). This obligation to warn leads some industry experts to believe that product labeling is the key factor in reduction of pharmaceutical liability risk.

Reliance on learned intermediaries also reduces risk since the ultimate decision to select and prescribe a prescription drug lies with the physician and not the patient. While permitted, advertising directed to either patients or physicians is tightly regulated and scrutinized. A drug may not be promoted to the extent that an "otherwise adequate warning becomes inadequate."[3] There are rigorous regulatory guidelines for packaging, promotional activities and events, educational forums, and even the sponsorship of research.

EMPOWERMENT OF THE FDA

The strict regulatory controls in place today evolved from two watershed events: deaths caused by lethal formulations of Elixir Sulfanilimide in the 1920s and tragic birth defects linked to the use of thalidomide in pregnant women in the 1960s. One hundred people died as a result of the use of diethylene glycol as a solvent for Elixir Sulfanilimide. Public reaction to this disaster led Congress to enact a "new drug" section in the 1938 Food, Drug, and Cosmetic Act. This empowered the FDA to evaluate the efficacy of all new drug formulations and to approve them as "safe for use" for indications specified on product labeling (Swazey, 1991).

Although never marketed in the United States, thalidomide was a popular sleeping drug introduced in 1957. It eventually was sold in 46 countries. In 1961 researchers discovered the association between thalidomide and phocomelia, or seal limbs, and other extreme congenital defects. By the end of 1961, thalidomide was withdrawn from most world markets. In the United States, the thalidomide tragedy led to the 1962 passage of the

Kefauver-Harris amendments to the Food, Drug, and Cosmetic Act, which tightened regulatory control over the safety of new drugs (Lasagna, 1991).

THE $359 MILLION ROAD TO MARKET

Today the drug development process typically spans 12 years and is costly and complex. Regulated development phases encompass laboratory testing; clinical studies of the pharmacologic profile of a new drug, its efficacy and tolerability by patients; and extensive clinical trials to study the effects of the drug in humans over specified periods of time. Pharmaceutical developers spend an average $359 million to bring a new drug to market, often for one limited application (U.S. Congress, Office of Technology Assessment, 1993). Very few drugs make the grade. In 1990 only 23 new drugs obtained clearance for marketing (Pharmaceutical Manufacturers Association, 1991). The attrition rate for new compounds is extraordinary. The Pharmaceutical Manufacturers Association estimates that only one in 5,000 new compounds completes the journey through the pipeline.

The drug development process begins with laboratory, or preclinical, testing of compounds that were discovered or acquired by a pharmaceutical manufacturer. During this time, which takes an average of three and one-half years, researchers seek to determine whether a compound is biologically active as well as safe. If preclinical tests yield promise in terms of human therapeutics, the pharmaceutical developer files an Investigational New Drug Application (IND) with the FDA before initiating clinical testing in volunteer patients. The IND provides exhaustive detail about the chemical, pharmacological, pharmaceutical, and toxicological properties of a new drug in the form in which it will be administered to patients.

Once the IND has been filed, approximately three years are spent in two phases of initial human clinical testing. In Phase I, 20 to 100 healthy people participate in studies where researchers observe the pharmacologic actions of the drug. These actions include the best tolerated dosing ranges; the manner in which the drug is absorbed, distributed, metabolized, and excreted; as well as the duration of the drug's therapeutic action. During Phase II, investigators use a battery of tests among 200 to 300 patients to obtain convincing evidence of the drug's medical benefits. Controlled tests are often used to measure the drug's effects against a placebo and are designed as open-label, or blinded, studies. Blinded studies are used to reduce subjective bias during analysis of a new compound. The total development time to this point averages six and one-half years. Once again, many compounds are dropped.

Drugs that survive the first two phases of human testing enter Phase III clinical testing to reconfirm whether the drug is effective and to identify any side effects that occur in statistically significant numbers of patients. Phase III studies are extensive and involve 1,000 to 3,000 volunteer patients in clinics and hospitals nationwide. The average length of time for Phase III testing is three years, which brings the time of development thus far to nine and one-half years. Products still fail and are terminated at this point.

REDUCING EXPOSURE OF PRODUCTS IN DEVELOPMENT

Measures to reduce exposure to liability are built into each research stage. Clinical investigators are indemnified. Institutional Review Boards are established at investigational sites to implement programs that monitor patient safety and ensure patient rights. Patients who volunteer for clinical trials are protected by stiff federal regulations and ethical standards. Federal regulation requires patients to be fully informed of the aims, methods, anticipated benefits, and potential hazards of a study before enrolling. Investigators must ensure that volunteers understand they are free to refuse to enter a study or to withdraw at any time. Written consent must be obtained from patients before they may participate. In the case of diminished ability, written consent is obtained from the patient's relatives or legal guardian. These measures appear to be effective: notwithstanding the recent experience with clinical trials of fialuridine as a treatment for hepatitis B, a review of case law suggests that manufacturers have not faced substantial litigation by clinical trial participants (Reisman, 1992).

Upon completion of Phase III trials, the developer files a New Drug Application (NDA) with the FDA to obtain a license to market the product for general use. In traditional paper form, the NDA can reach 90,000 pages and may actually fill an entire truck. In addition to providing the results of clinical testing, the NDA must include the suggested product labeling and drafts of advertising and promotional materials for FDA approval. While the FDA is required by law to review the NDA within six months, the average time for this process is two to three years.

THE HIGH PRICE OF REGULATION

This extended discussion of the drug development process is intended to prove a point: the safety and efficacy of a new drug have been rigorously evaluated prior to its entry into the market. Its medical benefits in relation to its potential risks have been calculated, weighed, and sanctioned. Clear, consistent product labeling has been developed to educate prescribers and encourage responsible administration of new drugs, thereby further reducing liability exposure. The developer has earned a reasonable assurance of protection against liability.

But at what price? Twelve years have passed. The pharmaceutical developer has spent $359 million without realizing a dime. And, liability is most likely to occur when an approved drug is used in populations that have not been substantially studied (Levine, 1993).

This is only one reason why clearance for marketing does not mark the end of the regulatory process. Post-marketing surveillance, or Phase IV, studies may take place to study emergence of new side effects or use of the drug in patient groups not previously explored. Additional studies may be conducted to compare a new drug with existing medications. Pharmaceutical developers may also wish to pursue new medical applications, new claims, or formulations of the drug beyond its approved indication. These studies may be conducted under the original, or new, IND as additional rigorously controlled Phase III trials, and may take up to four years. The high cost of additional trials often discourages manufacturers from seeking expanded indications.

Not even explored in this paper are the indirect costs to society of the regulatory process, including global competitiveness, the impact of the regulatory process on drug pricing, the limited access of critically ill patients to therapies in development, and reimbursement issues that arise when physicians prescribe drugs for off-label applications. Also not addressed is the disincentive to innovation caused by the fact that the regulatory process erodes patent protection. Another particularly critical issue is the need to ensure incentives for development of orphan drugs, or medicines for rare diseases with small patient populations.

IS REGULATION WORTH THE PRICE?

Are the enormous costs and painstaking pace of current regulation worth it? Does the grueling process pay off in reduced liability exposure for pharmaceutical products?

There is a widespread perception that the pharmaceutical industry is the victim of a nationwide litigation explosion. Indeed, a recent RAND study found that pharmaceuticals was a leading industry in federal liability suits, with a strong surge in case filings during the 1980s (Dungworth, 1988). But close analysis of the filings shows 60 percent of cases involved only two products, the Dalkon Shield and Bendectin. Further analysis shows that the Dalkon Shield and Bendectin rank second and third only to asbestos in terms of growth of federal product liability filings from 1974 to 1985 (General Accounting Office, 1988). These data suggest that instead of a litigation explosion, the pharmaceutical industry is vulnerable to concentrated clusters, or epidemics, of litigation. Cases involving Prozac, used to treat depression, and Halcion, for the treatment of insomnia, are current examples of this phenomenon.

CHILLING LESSONS OF BENDECTIN

The value of the tort liability system in situations where deception, fraud, or latent injury emerge is unquestioned. But the case of Bendectin is a cautionary tale.

Bendectin was the only prescription drug ever approved in the United States for the treatment of nausea and vomiting in pregnancy. Introduced in 1956, Bendectin was used in more than 30 million pregnancies. Beginning in 1969, assertions that Bendectin could produce congenital birth defects began to appear in scientific literature. While no sound scientific study ever proved a causal relationship between Bendectin and birth defects, and the FDA continued to affirm its safety, nearly 1,700 lawsuits were brought against the manufacturer (Pharmaceutical Manufacturers Association, 1993). The manufacturer won almost every case that went to court, but the price was too high. In 1983 the manufacturer voluntarily withdrew Bendectin from the marketplace because its $18 million annual cost for legal fees and insurance began to overwhelm its $20 million in annual sales (Viscusi and Moore, 1991). It is unlikely that any new drug will be developed to close this therapeutic gap.

DISPARITIES BETWEEN GOALS OF REGULATION AND LITIGATION

The Bendectin case illustrates disparities between the goals of the regulatory process and the tort liability system. The tort liability system is designed to discover risk and assign blame to a probable cause of injury *after it has happened*. On the other hand, the regulatory system seeks to predetermine potential adverse effects of a pharmaceutical product in order to *prevent or manage their occurrence*. In the case of Bendectin, assertions that it was a teratogen, or agent that causes birth defects, were based on the argument that *if* a pregnant women took Bendectin and *if* she gave birth to a deformed child, it was *possible* that Bendectin was the cause. This argument was designed to divorce each case from the backdrop of epidemiological data showing that chance alone could account for the incidence of 900,000 births of malformed babies among the 30 million women who took Bendectin while pregnant. It is interesting to note that the rate of birth defects has not declined in the United States since Bendectin was withdrawn. However, treatment for severe nausea during pregnancy now accounts for nearly $40 million of the nation's annual hospital bill (Pharmaceutical Manufacturers Association, 1993).

AREAS FOR TORT REFORM

Reforms are needed in many areas. One in particular—the issue of expert testimony—is indicative of the previously mentioned asymmetry between the objectives of regulation and those of product liability law. Courts have allowed expert witnesses to offer subjective or anecdotal testimony that is not based on sound, peer-reviewed scientific data. This practice of "junk science" has led to high jury awards in cases where no scientific evidence substantiated fault on the part of the manufacturer. The debate over "science in the courtroom" led to the 1993 Supreme Court ruling that federal judges must ensure that scientific evidence and testimony admitted in trials is "not only relevant, but reliable . . . that the methodology underlying the testimony is scientifically valid . . . that the theory or technique has been subjected to peer review" and that the "known or potential rate of error" of a particular scientific technique is considered.[4]

While the dismissal of "junk science" from federal courtrooms may eventually be good news for the industry, reform is needed in other areas of pharmaceutical product liability as well. The Pharmaceutical Manufacturers Association calls for the creation of a uniform federal tort liability system, as opposed to the current patchwork of state laws and the barring of punitive damages against manufacturers if they have already met the stringent requirements of the FDA. Other nations have diminished the lure of the industry's deep pockets and the impact of high jury awards by instituting social insurance schemes to redress injury. The national vaccine injury compensation program, which went into effect in the United States in 1988, resembles such measures and may signal the beginning of a trend.

A VIABLE EQUATION

It is true that the U.S. regulatory process and tort liability system greatly affect the cost of drug development and the length of time to market. The only real window for innovation is at the beginning of the 12-year regulatory process. The ultimate benefit of this process, however, is assurance of reasonably safe, effective prescription drugs.

Enormous strides in medical science have occurred in the past few decades despite the constraints of the system. Diseases have been eradicated. The prognosis for patients with diseases such as cancer and heart disease has dramatically improved. Biotechnology ushers in a new era of exciting possibilities, including the potential for safer, specifically targeted vaccines. The health of the industry appears to indicate that regulation, rational tort liability, and innovation represent a viable equation instead of an impossible triangle. Ongoing medical progress is telling evidence of a system that works.

NOTES

1. American Law Institute. 1965. Restatement (Second) of Torts. Sect. 402A, Comment k.
2. *McEwen v. Ortho Pharmaceutical Corp.* 1974. 270 Or. 375, 528. P. 2d 529.
3. *Stevens v. Parke Davis & Co.* 9 Cal. 3d 51, 67, 507 P.
4. *Daubert v. Merrell Dow Pharmaceuticals.* 1993 WL 224478 (U.S. June 28, 1993) (No. 92-102).

REFERENCES

Dungworth, T. 1988. Product Liability and the Business Sector: Litigation Trends in Federal Court. R-3668-ICJ. Santa Monica, Calif: RAND.

Fern, F. H., and W. M. Sichel. 1985. Failure to warn in drug cases: are punitive damages justifiable?" For the Defense 27:12–20.

General Accounting Office. 1988. Product liability; extent of "litigation explosion" questioned. Publication no. GAO/HRD-88-36 BR. Washington, D.C.: U.S. Government Printing Office.

Lasagna, L. 1991. The chilling effect of product liability on new drug development. Pages 334–359 in The Liability Maze: Impact of Liability Law on Safety and Innovation, P. W. Huber and R. E. Litan, eds. Washington, DC.: Brookings Institution.

Levine, C. 1993. Women as research subjects: New priorities, new questions. Pages 169–188 in Emerging Issues in Biomedical Policy: An Annual Review, R. H. Blank and A. L. Bonnickson, eds. Vol. 2. New York: Columbia University Press.

Pharmaceutical Manufacturers Association. 1991. Industry Issues Brief: The Drug Approval Process. Washington, D.C.: Pharmaceutical Manufacturers Association.

Pharmaceutical Manufacturers Association. 1993. Industry Issues Brief: Product Liability Reform. Tort Reform File. Washington, D.C.: Pharmaceutical Manufacturers Association.

Reisman, B. K. 1992. Products liability—What is the current situation and will it change (and how) when more women are included in studies? Presented at the Women in Clinical Trials Workshop of the FDA-Regulated Products Workshop, Food and Drug Law Institute, Washington, D.C.

Swazey, J. P. 1991. Prescription drug safety. Pages 291–333 in The Liability Maze: Impact of Liability Law on Safety and Innovation, P. W. Huber and R. E. Litan, eds. Washington, D.C.: Brookings Institution.

U.S. Congress, Office of Technology Assessment. 1993. Pharmaceutical R&D: Costs, risks and rewards. Publication no. OTA-H-522. Washington, D.C.: U.S. Government Printing Office.

Viscusi, W. K., and M. J. Moore. 1991. Rationalizing the relationship between product liability and innovation. Pages 105–126 in Tort Law and the Public Interest, P. H. Schuck, ed. New York: American Assembly, Columbia University.

THE SOCIAL, LEGAL, AND REGULATORY ENVIRONMENT

Insurance:
The Liability Messenger

DENNIS R. CONNOLLY

A t one point in the 1980s, there was a generalized "hard" insurance market. Insurance for a whole range of liability exposures, including product liability, was either very expensive or hard to come by, or both. Today, however, the insurance market is soft overall. In fact, the industry has the ability to write more insurance than it is currently writing. Even so, insurers are choosing not to divert this surplus capacity to cover certain classes of product liability risk, especially in industries such as pharmaceuticals, chemicals, automotives, and aviation.

Others have written eloquently about how costly U.S. product liability exposures are to consumers and manufacturers, principally by stifling innovation and competitiveness. The intent of this paper is to explain the process by which these risks become expensive to insure, or simply uninsurable. This is an important aspect of the product liability problem because, all too often, insurers are blamed for denying industry the affordable protection it needs as it goes about the business of innovating. In fact, both the blame and the solution lie elsewhere.

INSURANCE AS MESSENGER

To appreciate the role of the insurance system in managing corporate liabilities, it is helpful to think of the system as a message-bearer. Insurers send a message to insurance buyers whenever they set premium rates and establish the terms and conditions of coverage. When, for example, a company manufactures a product that causes injury to a consumer, compensates the consumer for his or her loss, and is itself indemnified under an applicable insurance policy, the insurer may then turn to the buyer and, in effect, say, "Your product has caused this problem, so we're going to raise your insurance premium."

Insurance buyers that fail to get the message and take the necessary precautions to reduce future losses will get harsher and harsher messages along these lines. When, for whatever reason, loss experience is not turned around, one of two things happens: either buyers reach a point where they cannot afford to pay for the coverage they need, or sellers reach a point where they cannot afford to extend the needed coverage at any price.

The fact is that today there is a strong sense among insurers that they cannot afford to insure parties for injuries and property damage caused by certain products and substances. For insurers, the known costs are too high and the potential costs are too uncertain.

UNDERWRITING: RESPONSIBLE RISK TAKING

To understand why the insurance system breaks down in certain cases, it is necessary to understand exactly how it works when it works well.

To provide insurance coverage, insurers must first engage in the art of underwriting. This is not a process of avoiding all loss. If it were possible to avoid all loss, insurance would not be necessary. Rather, when an insurance company places its assets at risk, it does so on the basis of reasonable predictions about how often losses can be expected to occur and what size claim payments they will be required to make.

Life insurance is probably as near to a science as the art of insurance underwriting comes. It is relatively easy to determine actuarially sound rates for particular individuals based on a great deal of experience with people of similar ages and lifestyles, for example. Underwriting in the property/casualty insurance industry is generally less precise.

As with life insurance, property/casualty insurers start with the available data. In determining rates for automobile liability coverage, insurers will first ask: What is this car's loss history? There are, for example, data supporting the contention that red automobiles are involved in more crashes than cars painted any other color. In fact, when it comes to automobile crashes, a database has been built up over time that can provide virtually any kind of statistical correlation desired.

When it comes to many product liability exposures, however, this is simply not the case. For new products, the data do not yet exist. And for complicated products, the available data do not provide a sufficiently strong basis for making credible predictions.

For such products, insurers use manuals produced by the Insurance Services Office, an industry organization for data collection and analysis.

Complicated products are designated with the letter "A," which means that the underwriter's own subjective judgment of the reliability of the product and the company is a key factor in the rate-making process.

This can be a difficult task, made more difficult by the fact that an underwriter must set a premium that encompasses a company's full product line. For instance, 3M has roughly 29,000 products. There is a good deal of art to figuring the appropriate premium for all 29,000 products.

In addition, an underwriter may make a reasonable assumption that a product poses little danger when used for its obvious purpose—but what happens when it is used in some other, more dangerous, context? One such case involves a bicycle bell technology that was being used as an altimeter in blimps to warn of rapidly decreasing altitude. When a blimp crashed, a lawsuit followed, naming the manufacturer of the bicycle bell as one of the defendants. But the bicycle bell manufacturer's insurer, in assessing the risk involved, had charged a premium based on the assumption that the bell would be used for bicycles, not blimps.

THE ROOTS OF UNINSURABILITY

Underwriting product liability generally entails accepting more uncertainty than, say, underwriting life insurance. Theoretically, the insurance purchaser, through appropriately priced premiums, protects the insurer from some, if not all, of this additional uncertainty. Unfortunately, for certain product liabilities, uncertainty rises to the point of wild unpredictability. As a result, no price may be high enough to compensate insurers for the potential losses they face.

In the pharmaceutical, chemical, automotive, and aviation industries, it is impossible to develop very useful risk assessment procedures because the number of variables is too great and the opportunity to isolate and analyze individual factors, and individual victims, does not exist.

But the usual difficulties of underwriting product risks are compounded manyfold by factors external to the nature and harmfulness of the products themselves. Probably the most pernicious source of added uncertainty is the U.S. tort liability system, which, it may be fairly said, has helped kill the messenger that is our insurance system.

LEGAL LIABILITY: SHORT-CIRCUITING THE INSURANCE MESSAGE

Product liability is, historically, a state matter. Thus, manufacturers must deal with 51 separate statutes and 51 separate bodies of case law interpreting those statutes. At the very least, it can be said that U.S. legislatures and courts send a profusion of mixed messages to business. But it is the liability principles built into these laws that most seriously interfere with the insurance message.

Liability Principles and Uninsurability

The business of commercial insurance involves spreading one party's risk among others. In determining the appropriate price for assuming the risk of a client with toxic or other product liability exposures, the insurer must assess risk on a case-by-case basis. However, joint-and-several liability, a staple of state product liability laws, is a major impediment to individual risk assessment. Under most state laws, individual policyholders may be forced to pay enormous court awards wholly out of proportion to their conduct. An underwriter who has reviewed the conduct of a particular insured and found it to be exemplary cannot simply develop a premium reflecting that fact. This is because the insured may become entangled in litigation involving actors whose conduct may be less commendable, but whose financial status is insufficient to bear their fair share of the liability burden. Thus, the ultimate losses incurred by individual insureds are highly unpredictable.

State laws can be insurer-unfriendly in other important ways:

- The use of strict liability magnifies risk, because companies that have done nothing wrong can still be held liable for court awards that are potentially life-threatening to them.
- Companies may be held liable for conduct which, at the time, complied with government standards. This is equivalent to changing the rules of the game after the bets have been made.
- The absence of caps on noneconomic, especially punitive, damages can result in awards hundreds of times as large as the economic loss caused.

Companies may also be held liable even if victims cannot identify the actual manufacturer responsible. For instance, in some states, a company's liability for DES injuries is dependent on its market share at the time the victim's mother took the drug.

Finally, in understandable sympathy for innocent victims, courts have been known to overreach. In one case, a dice manufacturer was sued when its dice allegedly emitted toxic fumes during a casino fire. While it was clear that the tiny volume of toxic fumes emitted by the dice did not contribute to any injury, that did not prevent the court, in a compassionate mode, from trying to arrange a nice compensation package for the injured party. When insurers are obliged to pay compensation for losses their policyholders had no hand in causing—not even accidentally—it is time for them to reconsider their underwriting practices.

Science, Pseudo-Science, and Uninsurability

For a number of reasons, underwriters must also concern themselves with the advance of science. First, new dangers may be found in products that were previously perceived as safe. Asbestos is a prime example of this. Policies of companies producing asbestos were underwritten in the 1940s. To have avoided the $1.5 billion settlement insurance companies will be paying to settle asbestos claims, the underwriters would have had to anticipate that later a firm causal link between asbestos and certain health problems would be established, and that the plaintiff's bar would be successful in the resulting lawsuits.

Second, the more scientifically advanced the product, the more uncertainty it is likely to engender in insurers. Precisely because it is such a departure from other products, it has no track record and thus provides no solid basis for predicting and pricing the risks involved.

Third, a product that can do a great deal of good for large numbers of people can sometimes do serious harm to a few. Vaccines are a prime example, especially vaccines that use live viruses that, in a relatively small number of cases, cause the disease they have been designed to protect against. If there is a vaccine for AIDS, it will prove simultaneously a boon for mankind and, perversely, a concern for product liability insurers.

Finally, when a product is improved, that should be a cause for celebration. But drug manufacturers have been known to celebrate too soon—unaware of the potential suits they face when they produce a new generation of drug that eliminates the side effects of the previous generation. Drug makers have been held liable for injuries caused by an old drug even though they lacked the scientific know-how to produce a drug free of side effects at the time.

While the above are examples of social goods posing certain difficulties for insurers, the onward march of science has a dark, insurer-unfriendly side, and that is the way our court system treats scientific evidence. In the United States, putatively harmful substances are well publicized and presumed to be lurking everywhere. As a result of this generalized paranoia, underwriters must also factor into their risk assessments the possibility that a court will impose a large bodily injury judgment on an insured based on what is either minority-supported evidence or, in some cases, just plain "junk science." Hence, uncertainty is piled on top of uncertainty.

The Litigation Crisis and Uninsurability

Finally, there is U.S. society's taste for litigation. Underwriters know that for every dollar spent in indemnifying a loss, 70 cents is spent on defense costs, and these costs are increasing 10 percent annually. Not surprisingly, in the areas where the plaintiff's bar has been most successful, insurance has virtually disappeared. Most pharmaceuticals, including vaccines, are "insured" through risk-funding mechanisms that are tantamount to self-insurance.

The tale of the swine flu vaccine is especially interesting. In 1975 Americans were urged to get vaccinated after an outbreak of swine flu. The insurance industry, anticipating the potential liability exposure for a drug being taken by such a large number of people, canceled the coverage of all four vaccine manufacturers. Congress then passed special legislation so that drug manufacturers would be liable only for manufacturing defects ("bad batch"); the government would take on the liability for any remaining exposure (design testing, for example).

During the time these decisions were being made, there was conflicting opinion on the extent of liability exposure involved. Many consumer groups and many in the plaintiff's bar said that the insurance industry was overreacting, that there would not be a great deal of litigation. Their arguments persuaded the government to assume some of the liability. The four companies paid an $8 million premium for coverage of manufacturing defects, but there was ultimately no litigation against them. However, the federal government did have to pay out more than $100 million to settle suits for the liability it assumed—despite the widespread assumption that it would not lose a penny—and, 20 years later, litigation is still pending.

One lesson this experience teaches is that it is very difficult to predict the extent of the liability that may result from a particular product. Another is that a joint effort by government, insurers, and manufacturers made the vaccination program workable: it could not have gone forward without such cooperation. There is much to be said about the pros and cons of private-public sector risk-sharing, but that is a topic for another paper.

With a legal liability system like this, it is not hard to see how the insurance message gets lost. For insurance to work as a message-bearer, it must be possible for insurance buyers to take measures to improve their loss experience. But the existence of a legal liability system that punishes good behavior as often as bad discourages insurers from staying in the game and leaves corporate risk managers with little to do but lobby their congressional representatives for national tort reform.

Other Causes of Uninsurability

The inflation of loss severity has also played a major role in souring insurers on certain product liability risks. The world has changed since the 1960s, when a judgment of $25,000 was considered a large loss. Several years later, $250,000 was considered a big loss. Today we counsel insurance companies to settle a class action lawsuit for silicone breast implants for $4.75 billion.

The increased frequency of lost lawsuits is also an issue. Product liability suits rank second behind medical malpractice in the number of nonvehicular cases resulting in million-dollar court awards. Toxic torts are especially onerous to insurers.

The continued growth in product liability awards and in statutory and judicial liabilities poses a major threat to a company's financial health. Insurers, who are in the business of *responsible* risk-taking, are understandably reluctant to sacrifice their own financial health to save their clients.

TORT REFORM: REVIVING THE MESSENGER

Not only does the out-of-control tort liability system undermine the insurance system, but the absence of insurance in turn undermines a principal aim of the tort liability system—to force wrongdoers to indemnify the innocent victim of their conduct. Companies without insurance are left totally exposed; when a loss occurs, they must pay it all themselves. The result is an increase in insolvent defendants, leading to an increase in uncompensated victims.

A world with dangerous products and toxic exposures is a perilous world indeed. The only thing worse is a world without insurance for the risks inherent in some classes of products or in innovation. Federal tort reform could, if it addresses the sources of insurer uncertainty, go a long way toward making uninsurable risks insurable again.

Junk Science in the Courtroom: The Impact on Innovation

PETER W. HUBER

There seem to be limits to the total uncertainty that any undertaking can bear, or so business experience suggests. Sellers of old products with established markets can sometimes shoulder the risk that attends operation within an unpredictable and sometimes capricious regime of liability. It is the new venture with the unfamiliar product that can least tolerate the extra measure of instability from a legal environment that does not provide predictable results. Insurance provides an antidote, if it can be obtained at a reasonable price.

In the past three decades, U.S. liability law has changed in many ways that have undermined insurance, and thus innovation. That effect was unintended. The architects of the modern U.S. liability system fully expected that insurance would continue to provide a broad financial umbrella over the expanding new tort system. But changes in liability law were implemented with little understanding of how insurance markets operate. The result has been a sharp increase in demand for liability insurance, but a marked decline in supply.

The most dramatic adjustment came in the early 1980s, with an insurance crisis eerily reminiscent of the endless gas lines during the Arab oil embargo. The insurance industry goes through regular cycles of "harder" and "softer" markets, and the market is currently softer (i.e., supply is greater) than it was some years ago. But the overall, long-term trends are not in serious dispute. Insuring high-litigation-risk products and services—obstetrics, contraceptives, light aircraft, vaccines, morning sickness drugs, and so on—is far more difficult today than it was 20 years ago.

A prudent insurer must know, first of all, just *who* it is insuring. It will write one contract for a driver with a history of drinking at the wheel, another (at a quite different price) for a well-run municipality, and yet another for a car manufacturer. But U.S. courts have steadily expanded principles of "joint" liability, which sweep together dozens, sometimes hundreds, of defendants in a single courtroom. "Several" liability then allows the full costs of an accident to be channeled to the wealthiest, inevitably the best-insured, defendant.

A second essential ingredient of intelligent insurance is accurate timing. How much risk an insurer is covering depends not only on when the policy begins and ends, which the insurer itself can control, but also on when legal claims are born and expire. In pricing a policy, the insurer counts on earning some investment income between the time when premiums are collected and the time when claims are paid. A reasonably predictable and quick clock on insurance claims also allows the insurer to identify better and worse risks. It will not do to sell liability insurance at a flat price to all comers for 20 years, and only then discover that some in the group are careful and others (as finally judged by the tort system) are grossly careless. But U.S. courts have been steadily dismantling principles of ripeness and limitation that once kept litigation on a fairly strict and predictable timetable. Lawsuits today can look forward and backward for decades, or even generations, with no one knowing which of the dozens of policies a customer might purchase during that period will then be called into play.

An insurer also needs a reliable yardstick for pricing injury. A policy speaks of accidents and such, but the final accounting is always in cash. A stable rate of exchange between injuries and dollars is therefore essential. The conversion is not hard to make when the injury is a broken leg that must be set or lost wages that must be replaced, which is why you can easily buy first-party health or disability insurance to cover such contingencies. But the rate of exchange is quite indeterminate for pain and suffering, loss of society, or criminal fines and penalties. For precisely this reason, first-party insurance never covers such things. In U.S. courts, as elsewhere, it used to be that tangible economic losses, like medical costs and lost wages, always accounted for most damages ultimately paid. In recent decades, however, there has been great expansion in U.S. awards for such things as punishment, pain, suffering, cancer phobia, and loss of society.

INSURANCE, RISK, AND THE ROOTS OF JUNK SCIENCE

A last and perhaps most essential ingredient of rational insurance is knowing just what risk is being covered. If an insurer sells a policy to a vaccine manufacturer, it must plan to cover the risks of vaccines, not the general health problems faced by young children. A policy priced to cover injuries caused by a spermicide cannot also cover birth defects originating in quite independent genetic accidents or a mother's drinking habit. Here too, in U.S. courts as elsewhere, the legal rules used to be fairly accommodating: claims of cause and effect were usually tested skeptically, and the courts declined to speculate about causes remote in time and place. But U.S. courts have gradually come to accept ignorance, or at least substantial uncertainty, about the risks of such things as toxic wastes as sufficient reason for setting the judicial machinery in motion. The upshot has been waves of litigation over "phantom risks,"[1] with marginal science peddled with impunity from the witness stand.

The legal disaster of the pertussis (whooping cough) vaccine unfolded in much the same way. The vaccine has virtually ended the 265,000 cases of pertussis and 7,500 pertussis-related deaths recorded in the years before 1949, the year the vaccine was first licensed. But a 1984 English study, serious and cautiously phrased in itself, suggested that the vaccine's use (extrapolated to the U.S. population) might be causing 25 cases a year of serious brain damage.[2] American lawyers responded with an avalanche of litigation, blaming the vaccine for brain damage, unexplained coma, Reye's syndrome, epilepsy, sudden infant death, and countless other afflictions.[3] Horrified pharmaceutical companies bailed out, and at one point it appeared that the last U.S. manufacturer of the product would be leaving the market.[4] More solid scientific evidence slowly accumulated.[5] Then, in March 1990, a report of a huge study of 230,000 children and 713,000 immunizations concluded that the vaccine had caused *no* serious neurological complications of any kind, and no deaths.[6] "It is time for the myth of pertussis vaccine encephalopathy to end," declared an editorial in the *Journal of the American Medical Association*. "We need to end this national nonsense."[7]

As experience now richly demonstrates, the incentives for lawyers today are simple and compelling. If the consensus in the scientific community is that a hazard is real and substantial, the trial bar will trumpet that consensus to support demands for compensation and punishment. If the consensus is that the hazard is imaginary or trivial, the bar will brush it aside and dredge up experts from the fringe to swear otherwise. Even when lawyers pursue certifiably real hazards, there will be a strong incentive to stretch claims to the margins of validity and beyond, to reach not just dangerous IUDs but also safe ones, not just serious exposures to asbestos but also trivial ones. If the law allows a lawyer to put just about anybody on the witness stand, she is going to search far and wide for any expertise, real or otherwise, that is congenial to her case.

It is a deep irony that the one place the law tolerates this sort of thing is in court. The professional seated in the witness box, alone among all other obstetricians, engineers, chemists, or pharmacologists, is above the rules. Or, to be more precise, he is often not subject to any rules at all. Malpractice by scientific and medical professionals is not only tolerated but encouraged, so long as it is solicited by lawyers themselves.

The law has always been ready enough to impose standards of competence on quacks outside the courtroom. Negligence law requires every doctor to "have and use the knowledge, skill and care ordinarily possessed and employed by members of the profession in good standing." If there are contending schools of thought in the profession, a malpractice defendant may be given the benefit of the doubt if he favors one school rather than another. But as a leading legal treatise hastens to add, this does not mean

that any quack, charlatan or crackpot can set himself up as a "school" and so apply his individual ideas without liability. A "school" must be a recognized one within a definite code of principles, and it must be the line of thought of a respectable minority of the profession.[8]

The designer of a product is likewise expected to have used "state-of-the-art" materials and technology.[9]

Often the law demands even more. In a famous 1932 decision concerning a tugboat called *The T. J. Hooper*, which (like most other tugs of the day) was not equipped with a radio, Judge Learned Hand sonorously declared that "in most cases, reasonable prudence is in fact common prudence; but strictly it is never its measure."[10] A court may thus require tugboat operators or doctors to surpass even accepted industry standards and consensus norms of the profession.[11]

From 1923 until the mid-1970s, the *Frye* rule made some attempt to hold expert witnesses to similar standards (*Frye v. United States*, 293 F. 1013 [D.C. Cir. 1923]). Certainly not to anything better than mainstream scientific norms, but the rule did at least refer to competent science as defined by the consensus views of a profession. Under *Frye*, the expert witness could report only learning that was "generally accepted" in his scientific discipline. Negligence, incompetence, irresponsibility, reckless disregard for professional standards, and every other variation on professional malpractice were as unacceptable on the witness stand as they were anywhere else.

One might suppose that this sort of symmetry would be a matter of fundamental fairness. If an obstetrician is to be judged guilty of malpractice, it will be on the say-so of some other doctor sitting in the witness box. A similar showdown between professionals decides every challenge to the design of a contraceptive or a Cuisinart. The jury must choose between yesterday's expert, who designed the morning-sickness drug or delivered the baby, and today's, who claims that the job was botched. Incredibly, many courts today enforce serious standards of professional competence only against defendants, not against their accusers.

The standards for medical witnesses are more biased still; the hermit clinician can usually testify to anything if he holds an M.D. and is willing to mumble some magic words about "reasonable medical certainty." Malpractice by mouth from the witness stand is thus not subject to any control at all. Any old résumé qualifies someone to be a witness, but only those who comply with the mainstream standards of professional medicine are good enough to escape liability.

The fringe theories and fanciful methods used to condemn experts in malpractice and product-design cases surface in other cases to explain disease. In the *Alcolac* litigation, for example, Zahalsky and Carnow, who testified for the plaintiff, based their chemical-AIDS conclusions on tests conducted with monoclonal antibodies.[12] The antibodies, however, were not approved by the Food and Drug Administration for any diagnostic purposes at all. Used in a clinic as a basis for treatment, such methods would have been actionable malpractice.

Our pursuit of incompetence among scientific and medical professionals is now often led by incompetents from the fringes of those same professions. Courts too eager to chase after distant and mysterious causes willingly attend to far-out, pseudoscientific mystics. In our eagerness to suppress inept, irresponsible, or fraudulent practice everywhere else, we have embraced inept, irresponsible, or fraudulent practice on the witness stand.

IMPACTS ON NEW TECHNOLOGIES

The impacts are felt throughout the economy, but most heavily on products, procedures, and technologies that are new and unfamiliar. Burt Rutan, the pioneering designer of the *Voyager*, used to sell construction plans for novel airplanes to do-it-yourselfers. In 1985, fearful of the lawsuits that would follow if a home-built plane crashed, he took the plans off the market. Meanwhile, the U.S. general aviation industry, once the unquestioned world leader, has almost ceased operations because of liability problems. In 1977 small-plane manufacturers in the United States paid a total of $24 million in liability claims. By 1985 their payout was $210 million. Companies like Beech, Cessna, and Piper sharply curtailed or completely suspended production; they quickly discovered that the new-model planes, carrying a 50 percent surcharge for liability insurance, could no longer compete with used planes already on the market. The market is moving steadily to manufacturers in Europe and South America.

For similar reasons, Monsanto decided in 1987 not to market a promising new filler and insulator made of calcium sodium metaphosphate. The material is almost certainly safer than asbestos, which it could help replace in brakes and gaskets. But safer is not good enough in today's climate of infectious litigation.

Liability fears have likewise caused the withdrawal of exotic drugs that the Food and Drug Administration deems safe and effective, including some for which no close substitute is known. Carl Djerassi, developer of the oral contraceptive, sees reform of the U.S. liability system as the most urgent priority in resurrecting the U.S. contraceptive industry.[13]

Some of the most regressive effects have been felt precisely where aggressive commercialization of new products is most urgently needed. U.S. liability today is highly, but often indiscriminately, contagious, which means that the introduction of new products is undercut the most in the markets already swept up in a hurricane of litigation—contraceptives, vaccines, obstetrical services, and light aircraft, for example. Yet it is often in such markets that new products are most urgently needed.

The commercialization of new products in the United States is thus discouraged on two sides. The availability of insurance in particular industries, and for particular products that have proved most attractive to litigants, has declined, and prices have increased, as the courts have adopted rules that discourage rational underwriting and systematic selection and differentiation of risks. At the same time, U.S. courts are effectively demanding that all goods and services come wrapped in a special-purpose insurance contract. The availability of insurance, most especially modestly priced insurance, depends largely on an accumulation of accident experience. That is something that established technologies always have and truly innovative ones never do. Insurance is easiest to find when a good has been used by many people for many years, so that the caprices of tort liability have been as far as possible washed out and the statistics of experience speak for themselves. Innovation necessarily starts without an established market, and so is often condemned to start without insurance as well. For the prudent businessperson, a start without insurance is often worse than no start at all.

LESS-THAN-OPTIMAL SOLUTIONS

How can a manufacturer commercialize a new but somewhat risky product absent affordable insurance, in a legal environment that always holds out the possibility for large verdicts, numerous lawsuits, and substantial expenses even in defending nonmeritorious claims?

One possibility, which was often suggested as a means for bringing the new French abortifacient RU-486 into the U.S. market, is to create a thinly capitalized, single-product firm, operate without insurance, siphon off profits quickly, and use the specter of immediate bankruptcy to deter litigation.[14] One small-plane manufacturer has attempted to operate in this way, and the option appears to be increasingly attractive for other small start-up companies. This is not much of solution, however. Most new products of any consequence require large, up-front investments to create a chain of distribution, publicize the product, and for such things as drugs, to shepherd the product through the very expensive and lengthy regulatory process.

A second possibility is to build a case-hardened litigation team and adopt a scorched-earth litigation policy to deter attack. Every large manufacturer does this to some extent, though the smart money seems to have concluded that this strategy is simply infeasible with certain classes of products used, for example, by pregnant women and young children. Tobacco companies are the only ones that have used this approach with almost complete success, and they have little in the way of new products to commercialize.

Yet another approach for the U.S. manufacturer is to establish a foreign base for the early stages of commercialization. This is by no means a complete solution, because it is becoming easier to pursue manufacturers back to U.S. shores, wherever products may be sold or used. Foreign manufacturers do have one important advantage. They can effectively limit their total liability for their U.S. operations by the cold-blooded expedient of keeping few of their assets in this country. Transnational recognition of judgment rules is still cumbersome enough to provide an effective shield against the wholesale export of U.S. liability verdicts to collect against the foreign-based assets of a foreign firm, no matter where its products may have been marketed.

THE *DAUBERT* RULING

In 1993 the U.S. Supreme Court took an important step in the right direction in ruling on the case of *Daubert v. Merrell Dow Pharmaceuticals*. The Court squarely held that district judges must play an important role as "gatekeepers," and admit only "reliable" and scientifically "relevant" evidence. The Court linked reliability directly to *scientific validity*, and stated that trial judges must make a preliminary determination that expert scientific testimony reflects "scientific knowledge." The two dissenting Justices reproached the majority for setting up judges in the role of "amateur scientists."

To be sure, *Daubert* will not immediately put an end to bad science in federal court. The Supreme Court emphasized the trial courts' "flexibility" in screening scientific evidence and did not designate failure to meet any particular criterion as automatically dispositive. This will undoubtedly tempt some judges who are uncomfortable in the gatekeeping role to engage in only cursory screening of scientific evidence. *Daubert* is a pragmatic, not dogmatic decision—and understandably so, given the wide practical sweep of the rules of evidence.

That said, the overall impact of *Daubert* should be favorable. The Court listed all the things that should be considered in screening scientific evidence. The opinion orders trial judges to determine whether a theory or technique can be (or has been) tested, and is therefore, as Karl Popper put it, "falsifiable." Peer review, the Court added, is an important, though not dispositive, factor to weigh. The Court also directed judges' attention to determining the known or potential rate of error of the technique in question, as well as the existence and maintenance of standards controlling the technique's operation. Moreover, general acceptance of the methods and theories at issue bear on the inquiry: the Court noted that "[w]idespread acceptance can be an important factor in ruling particular evidence admissible."

Most judges applying these factors will dismiss junk science claims before they reach trial. The trend toward stricter scrutiny of scientific evidence began in the late-1980s; in the aftermath of *Daubert* it will accelerate.

The four factors enumerated in *Daubert* as relevant to the admissibility of scientific evidence are likely to gather the lion's share of judicial attention. Attorneys faced with junk science claims, however, should recognize that the opinion gives them some easily overlooked but potentially powerful weapons.

The Court, for example, stated that proposed expert scientific testimony must have a valid scientific connection to the issue before the court to be admissible. As the Court emphasized, scientific validity for one purpose is not necessarily scientific validity for other purposes. High-dose animal studies, for example, arguably are relevant for risk research and could therefore be admissible in the context of litigation over Environmental Protection Agency regulations. The same studies, however, have questionable relevance for actually predicting harm to humans from low-dose exposure, much less in establishing that the particular substance at issue was more probably than not the cause of human injury. Such studies should be excluded if they are being used to prove causation.

The Court's focus on scientific relevance makes it clear that it is not sufficient that a claimant have a causation theory and an expert who is willing to testify at trial on the general subject at hand. Rather, the claimant must have an expert who can present testimony that will provide scientific support for the theory of causation being advanced.

Given that ruling, and the Court's emphasis on district judges as "gatekeepers," *Daubert* should encourage courts to issue appropriate pretrial orders to screen scientific evidence. An example of such an order has been routinely used by Federal District Judge J. Foy Guin, Jr.:

Within two weeks after the close of discovery, all parties are to file with the court summaries of the testimony to be given by any experts to be used at trial. Such summaries must include all conclusions and all reasoning supporting conclusions, including the scientific or professional theories or standards relied upon. These summaries may take the form of submitting copies of all reports submitted by the expert to the party hiring him, if they constitute a fair summary as aforesaid. At that time the court will determine the necessity of a court appointed expert pursuant to Rule 706, Federal Rules of Evidence. Failure on the part of any party to file said summaries will be grounds for forfeiture of that party's right to use expert testimony at trial.[15]

Junk science is offered on both sides of the courtroom aisle. Good science sometimes favors plaintiffs, sometimes defendants: what *Daubert* stands for is the proposition that judges have a major role to play in distinguishing accurate, reliable science from untested speculation, transparent error, or outright fraud.

No one can say whether the courts' single-minded pursuit of scientific truth will yield more universal justice than other approaches. Universal justice, justice not only for the parties but also for society as a whole, is, after all, a difficult thing to measure. But as the majority declares at the end of *Daubert*, judges, unlike scientists,

must resolve disputes finally and quickly. . . . Conjectures that are probably wrong are of little use . . . in the project of reaching a quick, final, and binding legal judgment—often of great consequence—about a particular set of events in the past.[16]

INSURANCE AS A MODERATOR OF RISK

How else might policy be changed to make liability a less serious obstacle to the commercialization of new products? One approach is to increase the supply of insurance. The other, to reduce the need for it.

Government has granted immunity, indemnification guarantees, or alternative insurance coverage to providers of various niche products and services, ranging from childhood vaccines to nuclear weapons. Similar approaches have been considered for "orphan drugs," which may be kept from market because the costs of insurance may exceed any potential profit. In other areas, however, immunities from liability have been cut back rather than expanded. Charities and the government itself used to enjoy blanket immunities from liability, but no longer do. In waiving its own absolute immunity from suit, however, the U.S. government carefully insisted on trial by judge, not jury, forbade punitive damages, and denied liability for any "discretionary function," the government equivalent of the judgment calls that a private company routinely makes in designing a new product.

More recently, and much less fruitfully, have been attempts to expand supply or curtail the cost of insurance by direct legislative decree. When insurers have canceled policies of their highest-risk customers, many states have passed laws limiting the insurer's freedom to cancel or refuse renewal. When coverage for particular lines has become entirely unavailable, some states have created joint underwriting associations, which force all private insurers doing any business in the state to offer coverage collectively, through the pool, for the otherwise uninsurable. When insurance prices have then risen still faster, several other states have turned to price controls. When insurers have attempted to flee, a few states have even challenged their right to depart. It seems likely, however, that such efforts will quite quickly reduce, rather than increase, the supply of insurance.

The second possible approach is to curtail demand, the need for insurance. This means changing the legal rules, so that cases are less frequent and judgments more modest. One of the blunter and more controversial approaches is to establish dollar caps on awards of such things as pain and suffering, or to limit punitive awards to some fixed multiple of the compensatory judgment. Many states have placed limits of one sort or another on joint liability, to reduce the increasingly magnetic properties of insurance, which otherwise cause the best insured to attract the most lawsuits. In addition to the 1993 *Daubert* ruling, there have been various other efforts to improve the quality of scientific testimony in the courts, so that factually unfounded claims cannot be used for their speculative value before a jury that may not understand the science in question. Other sensible guidelines for how the law can be changed to limit the chilling effects of open-ended liability can be found in the Federal Tort Claims Act, and in modern U.S. libel law.[17]

GETTING TO YES

In the end, the search must be for rules that allow society to say yes to new and better products, with the same conviction and force that an open-ended liability system can say no to old and inferior ones. In many areas of policy, the answers given depend largely on the question asked. For several decades, U.S. policymakers in the courts and elsewhere, have asked: What is unduly risky? And how can risk be deterred? But an equally important pair of questions is: What is acceptably safe? And how can safety be embraced? What indeed are our legal vehicles for getting to yes?

The first is private contract, which restores to individual buyers and sellers the power to make binding deals—deals addressing safety and risk along with other matters of risk—that will then be enforced, as written, in the courts. Current legal policy generally refuses to enforce such contracts, at least insofar as they address safety matters, on the assumption that consumers have unequal bargaining power, inadequate information, or simply too little patience to consider these matters.

But such rules could be changed. When we deal with the essentially private risks, of transportation or recreation for example, fair warning and conscious choice by the consumer must be made to count for much more than it does today. Not because individual choices will always be wise—they surely won't be—but because such a system at least permits positive choice and the acceptance of change.

A second vehicle for getting to yes is through the positive choice made by surrogates. Some safety matters will always remain too complex or far-reaching to be collapsed into the world of purely private agreement. Informed consent by the individual is never going to take care of such things as chemical waste disposal, mass vaccination, or central power generation. These are and obviously must remain matters to be delegated to expert agencies acting for the collective good. But if they are to be useful at all, agents must be able to buy as well as to sell. For safety agencies, this means not just rejecting bad safety choices, but also embracing good ones. Yet the long-standing rule, still rigidly applied in U.S. courts, is that even the most complete conformity to applicable regulations is no shield against liability.

The courts should be strongly encouraged to respect the risk and safety choices made by expert agencies. It is politically unrealistic to propose that liability be cut off entirely for any activity expressly approved by a qualified regulatory agency. But liability could at least be firmly curtailed in such circumstances. Some states, like New Jersey, have already adopted reform along these lines for pharmaceuticals and other products comprehensively regulated by federal agencies. At the very minimum, complete compliance with a comprehensive licensing order should provide liability protection against punitive damages—awards that have a strong, depressive effect on innovative experiment and change. It has always been true that ignorance of the law is no excuse. At present, knowledge of the law is no excuse either. It should be.

CONCLUSION

We may end, then, where we began, with the issue of insurance. Insurance remains the best and most robust market instrument for making the random predictable. It is possible to insure against misadventure in the courts, just as it is possible to insure against wind and wave on the high seas. But only up to a point. When the legal rules are such that insurance itself attracts litigation, when the rules change too quickly for actuarial assessment, and when systematic biases creep in against new technologies and innovative undertakings, insurance will rise in price and narrow in coverage. No other country in the world has gone through liability insurance shocks like those experienced in the United States. The reason is that no other country has transformed its liability system so completely and abruptly. It is within the direct power of U.S. courts to restore a climate that is healthy for private insurance markets. The courts thus have a very direct role to play in shaping policies to renew the spirit of enterprise in the United States, and recreate a social climate that welcomes, indeed encourages, the commercialization of new products.

NOTES

This paper is adapted from P. W. Huber, *Galileo's Revenge: Junk Science in the Courtroom* (Basic Books 1991); *Phantom Risk: Scientific Inference and the Law* (K. R. Foster, D. Bernstein, P. W. Huber, eds. MIT Press 1993); and D. Bernstein and P. Huber, *'Daubert' Plaintiffs Won Technical Battle But Plainly Lost the War*, 21 Product Safety & Liability Reporter 16 (BNA Special Report Summer/Fall 1993).

For the views of a plaintiff's lawyer who disagrees with me on virtually everything, see Chesebro, *Galileo's Retort: Peter Huber's Junk Scholarship*, The 42 Am. Univ. L. Rev. 1637 (Summer 1993).

1. See generally *Phantom Risk: Scientific Inference and the Law*, (K. Foster, D. Bernstein, and P. Huber, eds., MIT Press 1993).

2. See Hinmon and Koplan, *Pertussis and Pertussis Vaccine: Reanalysis of Benefits, Risks, and Costs*, 251 Journal of the American Medical Association 3,109 (1984).

3. See H. Coulter and B. Fisher, *DPT: A Shot in the Dark* (1984); Engelberg, *Vaccine: Assessing Risks and Benefits*, New York Times, December 19, 1984, sec. 3, p. 1.

4. See Russell, *Firm Ceases Making Vaccine*, Washington Post, June 19, 1984, p. A1; *The Cost of Ignoring Vaccine Victims*, New York Times, October 15, 1984, sec. 1, p. 18; Boffey, *Vaccine Liability Threatens Supplies*, New York Times, June 26, 1984, sec. 3, p. 1; Engelberg, *Vaccines*, p. 21; Engelberg, *Official Explains Gaffe on Vaccine Shortage*, New York Times, December 15, 1984, sec. 1, p. 10.

5. E.g., MacRae, *Epidemiology, Encephalopathy, and Pertussis Vaccine*, in FEMS-Symposium Pertussis: Proceedings of the Conference Organized by the Society of Microbiology and Epidemiology of the GDR, April 20-22, 1988, Berlin; Stephenson, *Pertussis Vaccine on Trial: Science vs the Law (High Court of London)* in ibid.; Walker et al., *Neurologic Events Following Diphtheria-Tetanus-Pertussis Immunization*, 81 Pediatrics 345-49 (1988); Shields et al., *Relationship of Pertussis Immunization to the Onset of Neurologic Disorders*, 113 Journal of Pediatrics 801-805 (1988).

6. Griffin et al., *Risk of Seizures and Encephalopathy After Immunization with the Diphtheria-Tetanus-Pertussis Vaccine*, 263 Journal of the American Medical Association 1,641-1,645 (March 23-30, 1990); see also Kolata, *Whooping Cough Vaccine Found Not to Be Linked to Brain Damage*, New York Times, March 23, sec. 1, p. 19.

7. Cherry, *"Pertussis Vaccine Encephalopathy": It Is Time to Recognize It as the Myth That It Is*, 263 Journal of the American Medical Association 1679-80 (March 23-30, 1990).

8. W. Keeton, D. Dobbs, R. Keeton, and D. Owen, Prosser and Keeton on Torts 187 (5th ed., 1984).

9. See, e.g., *Toliver v. General Motors Corp.*, 482 So.2d 213 (Miss. 1985).

10. The T. J. Hooper, 60 F.2d 737, 740 (2d Cir.), *cert. denied*, 287 U.S. 662 (1932).

11. E.g., *Johnson v. Hannibal Mower Corp.*, 679 S.W.2d 884 (Mo. App. 1984).

12. Alcolac, Inc., manufactured specialty chemicals for soaps and cosmetics at a plant. Pollution from that plant was said to have damaged the immune systems of families who lived nearby. See *Elam v. Alcolac, Inc.*, 765 S.W. 2d 42, 212-14 (Missouri. App. 1988).

13. C. Djerassi, *The Bitter Pill*, Science, Vol. 245, p. 356, 357 (28 July 1989).

14. RU-486's manufacturer has donated all U.S. patent rights to the Population Council, a nonprofit organization based in Manhattan. See L. Garrett, *Abortion Pill Test: Backers Say Prescriptions in U.S. Possible Next Year*, Newsday, May 17, 1994, at A05.

15. Many state courts issue so-called *Lone Pine* orders to similar effect. *Lore v. Lone Pine Corp.*, No. L-033706-85 (N.J. Super. Ct. Nov. 18, 1986).

16. Slip op. at 16-17.

17. The Federal Tort Claims Act governs claims by citizens against the federal government. The act diverges sharply from ordinary tort law in both the procedures it establishes (e.g., trial by judge, not jury) and the substantive standards it prescribes (on such things as punitive damages, discretionary calls, and so on). U.S. libel law likewise sets standards of liability so as to minimize the chilling effect of tort litigation on the press, and requires close review of jury verdicts by appellate courts.

Product Safety Regulation and the Law of Torts

SUSAN ROSE-ACKERMAN

Despite many convincing anecdotes, tort law's impact on technological innovation is inconclusive. The best studies suggest that
the relationship, if one exists, is complex and multidimensional.[1]
Although real-world anecdotes are suggestive, they are unreliable as a
basis for making policy. One hears many stories, for example, about the
negative impact of tort suits on the practice of medicine. However, research has shown that only one in 10 injuries attributable to medical malpractice results in a lawsuit.[2] This finding undercuts the contention that
an avalanche of frivolous malpractice suits has made doctors excessively
cautious. Similarly, the more extreme claims linking product liability suits
to lowered levels of research and development appear to have no basis in
fact. One study has even found that at low to moderate levels of expected
liability, increased liability costs encourage research and development.[3]
Only at very high liability levels is the effect negative.[4] This result is quite
plausible since increases in expected liability costs could induce spending
on innovations to improve safety.[5]

Nevertheless, even if the alarm expressed by some observers is over-
stated, tort law may still be creating inefficient incentives for product innovation. If we believe that the market does not create efficient incentives
to produce safe and healthy products, perhaps direct regulation is superior to tort law. If tort law will remain a fixture of the American legal landscape, perhaps it can be redesigned to complement the regulatory system
rather than work in opposition.

TORTS VERSUS STATUTES

Both tort law and statutory law have regulatory effects. This paper will address the question of how tort law and regulation by statute should fit together. While some critics contend that the size of jury awards implies that the tort system should be entirely abandoned as a way of regulating product quality, that is an unrealistic proposal. Nevertheless, one can distinguish the relative efficiency of tort law and statutory law in alternative settings. Different situations may require reliance on one system over the other, as the following examples suggest.[6]

Statutory law works best in the following circumstances:

1. If the harm is very diffuse, with many people harmed in a small way. No one has much of an incentive to sue individually, and even though class action suits are an option, they are not always effective.

2. If damages are imposed on large numbers of people. In these cases, the tort system, which operates on an ex post and case-by-case basis, may be less effective than an ex ante regulatory system. The latter approach facilitates economies of scale and conserves on information in cases of individualized but similar harms. Such problems are most effectively controlled through a standard-setting process at the government level.

3. When the damages cannot be tied to a single, identifiable source. For example, if polluting smokestacks are causing many people to suffer, there is nothing to be gained by reducing the problem to a set of disputes between particular individuals and particular smokestacks. Pollution damages are a statistical problem, and it is a waste of resources to try to control damages through a set of individualized decisions in the tort system.

4. If the companies causing the injuries are too poor to pay for the harm they cause.

In contrast, the tort system is more effective for regulating very low probability events because it may be administratively cheaper. Developing, enacting, and enforcing regulations is a time-consuming and expensive process, and one does not want to burden the system by over-regulation.

In theory, the relative importance of these factors should help one choose between regulating through the incentives provided by the tort system or through an ex ante statutory system. However, in practice, in areas like toxic torts, product liability, and medical malpractice, the line between the two systems has blurred. The courts have innovated by making themselves into little regulatory agencies and setting up institutions to administer their judgments. This is a trend that has little to recommend it.

Courts are not very good at acting like agencies. When one sees this happening, that is an argument for establishing regulatory institutions.

TORTS AND STATUTES AS COMPLEMENTS

Given the existing interconnections and overlaps, is there some way to make the tort system and the regulatory system complementary, rather than competitive? Three possibilities exist.

First, when various dangers exist that have not been regulated, the tort system provides a stop-gap measure pending the passage of legislation. Thus in *Larsen v. General Motors* (391 F.2d 495, 506 [8th Cir. 1968]) a federal court of appeals holds that "[t]he common law standard of [reasonableness] . . . can at least serve the needs of our society until the legislature imposes higher standards." The court recognizes that a problem exists and acknowledges that there has not been a systematic approach from the relevant regulatory body. Thus, the courts are filling in the gaps, albeit in an imperfect way.

Second, statutes are often treated as baselines in tort suits. The federal courts have taken this approach in product and occupational health and safety cases. They have ruled, for example, that the Food, Drug, and Cosmetic Act sets only minima and does not preempt tort suits for damages (*Abbot v. American Cyanamid Co.*, 844 F.2d 1108 [4th Cir. 1988]). Similarly, tort suits involving automobile design (15 U.S.C. § 1397(c) and e.g. *Sours v. General Motors Corp.* 717 F.2d 1511, 1516-1517 [6th Cir. 1983]) or exposure to nuclear materials (*Silkwood v. Kerr-McGee*, 464 U.S. 238 [1984]) are not preempted by regulatory statutes.[7]

In the occupational safety and health area, a baseline statute could designate the most obvious risks, the risks that everyone would agree to eliminate. Then bargaining between workers and management could establish higher standards in particular workplaces or industries. The tort system, then, would make it possible for people to argue that the standards should be higher in particular cases. In the context of a statute that was explicitly a baseline, the tort system would accommodate special cases, given the fact that the world is more diffuse than the regulation indicates. By the same token, a manufacturer should be able to argue that it should not be held to as high a standard as others because of the particular way its product is used. The court system could provide for these exceptions.

The third possibility for complementarity is a tort system that acts as a supplemental enforcement device. This is possible when the negligence rule in torts is equal to or less stringent than the regulatory standard. It is also possible in a pure, strict (or absolute) liability system that imposes liability whenever a product-linked injury occurs. When the tort system acts as such a compensation system, or when the negligence standard is no higher than the regulatory standard, a company that does not meet either the regulatory or the negligence standard can be held liable and required to pay damages. If the company meets the regulatory standard, it would not be held liable.[8]

Obviously, problems will arise if the tort system, in practice, imposes standards that are higher than those of the regulatory agency. By doing so, it undermines the notion that the regulatory system sets standards where benefits are balanced against costs. Under such conditions the two systems would work at cross-purposes. The conflict between torts and regulation would be exacerbated by the existence of punitive damages.[9]

Labeling law provides an example of how tort law can complement direct regulation. In 1993 two circuit courts ruled on federal preemption of tort claims, one under the pesticide statute and the other under the hazardous substance law.[10] The statutes contain very similar language. Both clearly preempt state statutes that try to impose different labeling requirements. The laws do not directly address the issue of tort suits.

The cases were decided by the fourth and the eleventh circuits. Under the principles articulated here, the fourth circuit made the correct ruling, and the eleventh circuit did not. The eleventh circuit, in a case involving the pesticide law, simply concluded that lawsuits challenging the adequacy of labels were preempted.[11] The fourth circuit took a more nuanced approach. In a case dealing with an exploding paint thinner, the court found that the labels conformed with the law. It then went on to argue that preemption only applies to suits claiming that a label, which complies with the federal standard, should have been stronger. A tort suit is permitted, however, if one can demonstrate that the label did not conform to federal standards.[12] In this way, the tort system operates as an enforcement device. It supplements the limited resources of public regulatory agencies without undermining statutory goals.

REGULATORY REFORM

Commentators have long urged legislators and regulatory agencies to charge fees set to reflect the risks created by regulated firms and to establish performance-based standards. Incentive-based reforms allocate regulatory costs to those who can bear them most efficiently, encourage firms to search for innovative ways to reduce harms, and force producers to reflect the risks they impose on society.

Such reforms, however, could be undermined by a poorly informed judiciary. If courts equate regulation with standard setting, then they may treat only command-and-control regulation as behaviorally significant. In a recent case, for example, the Superfund law was described as "not a regulatory standard-setting statute" because polluters pay for the cost of abating hazardous wastes "through tax and reimbursement liability" (*State of New York v. Shore Realty*, 759 F.2d 1032, 1041 [2d Cir. 1985]).

Incentive schemes require a fundamental rethinking of the relationship between tort law and statutory law. Following the conventional wisdom of economists and policy analysts, regulators have begun to use incentives and subsidies to affect behavior in lieu of command-and-control standards. The Environmental Protection Agency has experimented with "bubbles," "offsets," and "banking." The 1990 Amendments to the Clean Air Act seek to control acid rain through a system of tradeable pollution rights (Clean Air Act of 1990, §§ 403–405). Similar proposals exist to pay workers to use protective devices under the Occupational Safety and Health Act and to establish marketable rights for water pollution.

How should courts handle claims by defendants that incentive-based regulatory statutes preempt tort actions? Judges who view regulation as confined to standard setting might allow tort actions on the ground that these statutes are not "regulatory" because they do not establish uniform standards but "only" create incentives. Yet the argument for preemption of tort law is even stronger in the case of incentive-based regulations than in the case of command-and-control regulation. With standard setting based on either technology or performance, tort actions can complement regulatory agency activity if agency enforcement is not comprehensive or if the fines levied bear little relationship to damages. In contrast, a well-designed incentive system signals to a firm the social costs of its activities. A fee system resembles a tort liability system: No fixed standards are set, but firms respond to the cost of damages. The regulated entity must purchase the right to impose social costs in the same way that a tort judgment requires payment for harms. The main difference is the comprehensiveness of a fee schedule, which the state sets so that all firms are covered. A firm's liability does not depend on the contingency of private litigation and jury damage awards.

If fee schedules are set to reflect the social costs of the regulated firm's activities, then tort actions would be redundant at best and counterproductive at worst. Tort judgments would undermine such a regulatory scheme, especially if courts applied a strict liability standard, the type of standard that some judges have found least "regulatory" (*Silkwood v. Kerr-McGee*, at 276 n.3). Thus, incentive-based statutes should include a provision clearly preempting tort actions. For example, if the Environmental Protection Agency charges effluent fees, those damaged by the discharges that occur should not be able to sue since this would create inefficient care-taking incentives on the margin.

The only role for lawsuits by private individuals would be to force the *agency* to enforce its own rules; such suits might permit private recovery of damages for harm caused by lax enforcement. For example, the Consumer

Product Safety Act permits suits for damages against firms that violate agency rules. In situations where the damages are too diffuse to motivate private litigation, the recovery could be some multiple of fees that the agency could have exacted and could be paid to the Treasury with the public interest litigant recovering legal fees. Thus, although ordinary tort actions would be preempted, certain specialized private remedies might supplement agency enforcement just as tort actions do which use regulatory standards as the standard of negligence.

COMPENSATION

Tort law provides more than a set of regulatory incentives; behavior modification is not its only legitimate function. It is also a compensation system triggered by victims' complaints. If a regulatory statute bars private tort actions, those who were previously able to sue for damages will be disadvantaged, a result courts seem reluctant to permit. In finding that Karen Silkwood could sue for punitive damages in state court despite a federal statute that preempted state regulation of the nuclear industry, the Supreme Court noted that the statute did not provide for compensation and stated that "It is difficult to believe that Congress would, without comment, remove all means of judicial recourse for those injured by illegal conduct" (*Silkwood v. Kerr-McGee* at 251). If compensation of victims is not addressed by a purely regulatory statute yet remains a policy goal, conflict may arise between the statute and tort law. Compensation-oriented courts may apply conventional tort doctrines that are at cross purposes with regulatory policies.

We need to focus on situations where regulatory policies conflict with a compensation-oriented tort law. Where truly innocent victims exist, denying compensation to those who formerly could bring damage actions may be unjust and unwise. Yet retaining conventional tort actions in the face of regulatory statutes can undermine the behavioral impact of statutes. Other solutions must be found to the problem of providing compensation.

If the victims are numerous and their losses fall into broad, easily identified categories, such as lost limbs or particular types of cancers, then the compensation goal could be served by direct subsidy programs similar to workers' compensation or the black lung compensation program. In contrast, if the victims are few in number and their problems are idiosyncratic, the law should either permit private rights of action for damages analogous to those permitted under the Consumer Product Safety Act and the Comprehensive Environmental Response, Compensation, and Liability Act (Superfund),[13] or it should allow tort actions under strict liability principles solely as a means of achieving compensation.

CONCLUSIONS

The tort system deals inadequately with problems that do not fit easily into traditional tort categories, problems such as latent cancer risks and harms with attenuated chains of causation. The innovations that the courts have developed to manage class actions and consolidate cases are transforming the courts into quasi-regulatory agencies. Real agencies are likely to perform better than awkward judicial hybrids that have many of the disadvantages of both forms.

If Congress reforms the regulatory system to rely more heavily on incentive schemes, the judicial role should become even more modest. Under incentive schemes that require firms to pay for the damage they cause, statutes should preempt tort actions in order to avoid overdeterrence. For programs affecting many people, compensation should be effected through a separate system of social insurance. Private lawsuits would be permitted under the statute only to compel regulated entities to comply with existing regulatory standards.

But in policy areas that have not yet been reformed, a limited role remains for tort law or, at least, for private causes of action embedded in statutory schemes. Negligence law can be complementary to command-and-control regulation if it adopts the agency's standard not just as a minimum but as the measure of due care. Conversely, a true, strict (or absolute) liability regime would obviate a judicial risk-benefit calculation; only a determination of causation would be required. The choice between negligence and strict liability should then depend on how society evaluates the importance of giving victims an incentive to take care versus the distributive effects of initially shifting all losses to injurers.

An efficiently operating system of tort and regulatory law might indeed affect the research choices of business firms, but that would be a result to applaud, not condemn. If manufacturers are induced to take into account the costs imposed by their products on society, this will give them an incentive to make appropriate research and development choices. Innovation will not be discouraged, but it may be redirected. The current mixture of tort law and direct statutory regulation, however, does not appear to conform to the economic ideal. Policymakers should, however, seek a more consistent system, not impose artificial limits on either tort judgments or regulatory initiatives.

NOTES

Portions of these remarks are derived from Susan Rose-Ackerman, "Regulation and the Law of Torts," *American Economic Review-Papers and Proceedings*, 81:54-58 (May 1991). A more extended discussion can be found in Susan Rose-Ackerman, *Rethinking the Progressive Agenda: The Reform of the American Regulatory State* New York: Free Press, 1992, Chapter 8, pp. 118–131.

1. See W. Kip Viscusi and Michael J. Moore, "Rationalizing the Relationship between Product Liability and Innovation," and Robert Litan, "The Liability Explosion and American Trade Performance: Myths and Realities," both in Peter Schuck, ed. *Tort Law and the Public Interest: Competition, Innovation, and Consumer Welfare*, N.Y.: Norton, 1991, pp. 105-150. W. Kip Viscusi and Michael J. Moore, "Product Liability, Research and Development, and Innovation," *Journal of Political Economy*, 101:161-184 (1993).

2. Patricia Danzon found that roughly one in 126 patients admitted to California hospitals in 1974 suffered an injury due to negligent medical care. Of these no more than one in 10 filed a claim and only 40 percent of these claims resulted in payment to the patient. Patricia Danzon, "Malpractice Liability: Is the Grass on the Other Side Greener?" in Peter Schuck, eds. *Tort Law and the Public Interest: Competition, Innovation, and the Consumer Welfare*, N.Y.: Norton, 1991, pp. 176-204 at 183.

3. W. Kip Viscusi and Michael J. Moore, "Product Liability, Research and Development, and Innovation," *Journal of Political Economy*, 101:161-184 (1993).

4. Eleven three-digit SIC industries were in the high liability group. One of them was the composition goods industry which includes asbestos manufacturers. Another was miscellaneous chemicals, a group including manufacturers of battery acid, fireworks, jet fuel igniters, and pyrotechnic ammunition. Id. at 181.

5. Id. at p. 167.

6. The discussion in this section is derived from Steven Shavell, "Liability versus Other Approaches to the Control of Risk," in Shavell, *Economic Analysis of Accident Law*, Cambridge, Mass.: Harvard University Press, 1987, pp. 277-290.

7. In contrast, *Wood v. General Motors Corp.* (865 F.2d 395, 402 [1st Cir. 1988]) held that common law actions seeking to hold manufacturers liable for failing to install airbags were *not* permitted. Airbags are a special case because the agency expressly permitted automotive firms to select an alternative to airbags.

8. A study of compliance with Occupational Safety and Health Act standards by the custom woodworking industry found high levels of compliance despite weak agency enforcement. The author speculates that one reason may be the fear that violation of an OSHA standard would leave the employer open to higher liabilities from tort judgments, workers compensation premiums, or insurance ratings. David Weil, "If OSHA Is So Bad, Why Is Compliance So Good?", draft manuscript, Boston University, Boston, Mass., 1993, pp. 31-32.

9. For a fuller discussion, see Susan Rose-Ackerman, *Rethinking the Progressive Agenda: The Reform of the American Regulatory State*, New York: Free Press, 1992, Chapter 8, p. 127.

10. *Papas v. Upjohn Co.*, 985 F.2d 516 (11th Cir. 1993); *Moss v. Parks Corp.*, 985 F.2d 736 (4th Cir. 1993), cert. denied 113 S.Ct. 2999 (1993).

11. 985 F.2d 516 at 517.

12. 985 F.2d 736 at 739-741.

13. Under the CERCLA, private individuals can sue generators of hazardous wastes for cleanup costs (but not for personal injuries) even if the government has taken no action against the waste generator.

The Inconvenient Public: Behavioral Research Approaches to Reducing Product Liability Risks

BARUCH FISCHHOFF AND JON F. MERZ

here would not be any product liability suits if there were not any people involved with engineered systems. Unfortunately, people are everywhere, and they sometimes make mistakes—as consumers, operators, and patients. They misunderstand instructions, overlook warning labels, and employ equipment for inappropriate purposes.

In many cases, they realize that any ensuing misfortune is clearly their own fault, as when they have been drinking or using illicit drugs. Often, though, their natural response is to blame someone else for what went wrong. In psychological terms, there are both *cognitive* and *motivational* reasons for this tendency. Cognitively, injured parties see themselves as having been doing something that seemed sensible at the time, and not looking for trouble. As a result, any accident comes as a surprise. If it was to be avoided, then someone else needed to provide the missing expertise and protection. Motivationally, no one wants to feel responsible for an accident. That just adds insult to injury, as well as forfeiting the chance for emotional and financial redress.

Of course, the natural targets for such blame are those who created and distributed the product or equipment involved in an accident. They could have improved the design to prevent accidents. They should have done more to ensure that the product would not fail in expected use. They could have provided better warnings and instructions in how to use the product. They could have sacrificed profits or forgone sales, rather than let users bear (what now seem to have been) unacceptable risks.

It is equally natural for producers and distributors to shift the blame back to the user. Cognitively, the wisdom of hindsight makes it obvious to

159

them what the user should have done or seen in order to avoid an accident. They remember all the care that was taken in the design process. They see no ambiguity in the instructions and accompanying warnings. They would not have dreamed of using the system or product in the way that led to the accident. Motivationally, no one wants to be responsible for another's misfortune, even where there are no financial consequences. No one is in the business of hurting people.

To the extent that this description is accurate, it depicts an unhappy situation. Accidents keep happening, while each side blames the other. At the extreme, injured users may translate their grievances into lawsuits, while the producers and distributors fume about the irresponsible public. There is, of course, a steady supply of lawyers, politicians, and pundits ready to fan these frustrations. In the short run, it can be reinforcing to hear about the other party's venality or incompetence. In the long run, though, such sweeping claims merely reinforce prejudices and obscure the opportunities for progress.

In this light, technical innovation is threatened not just by an unthinking public, but also by an unthinking attitude toward the public. Few people in the technical community have any significant training in the behavioral sciences. As a result, it is hard for them to make sense of the behavior that they see or to devise creative improvements in design. They may have been drawn to engineering because it promised greater predictability than did dealing with fallible people. They may be reluctant to acknowledge the limits of their expertise or to include new kinds of expertise in already complex design processes.

This paper will analyze the opportunities for incorporating scientific knowledge about one aspect of human behavior in the product design and management process: how people understand the risks of the products they use. It will look at both quantitative understanding, regarding the magnitude of risks and benefits, and qualitative understanding, regarding how risks are created and controlled.

Quantitative understanding is essential if people are to realize what risks they are taking, decide whether those risks are justified by the accompanying benefits, and confer informed consent for bearing them. Qualitative understanding is essential to using products in ways that achieve minimal risk levels, to recognizing when things are going wrong, and to responding to surprises.

After presenting some of the empirical and analytical procedures for assessing and improving these kinds of understanding, this paper will consider the extent of their possible contribution to product safety and innovation. Its goal is to encourage attention to these issues in the product stewardship process.

ARENAS FOR RISK PERCEPTIONS

Although technical experts have the luxury of specializing in the management of particular risks, members of the general public do not. They face too many risks in their lives to acquire detailed knowledge of more than a minute portion of those risks. Their risks include affairs of the heart, ballot box, and pocketbook, as well matters of health and safety. Even in matters of physical welfare and survival, the list of concerns can be very long. Table 1 provides an illustrative list of situations in which the risk perceptions of individuals have consequences.

The enormous range of risks creates both challenges and opportunities for the manufacturers and distributors of potentially hazardous products. On the one hand, they must fight to divert a portion of the public's scarce attention to the potential risks of their products. In so doing, they may imperil their financial security by diverting attention from the benefits of

TABLE 1 Arenas for Risk Perception

Workplace
- On-the-job safety
- Right-to-know laws
- Workers' compensation

Neighborhood
- Rumors
- Emergency response
- Community right-to-know
- Siting

Courts
- Informed consent
- Risk-utility analysis
- Psychological stress

Regulation
- Agenda setting
- Safety standards
- Local initiatives

Industry
- Innovation
- Public relations
- Insurance
- Product differentiation (by safety)

those products or by making their products seem riskier than other (possibly competing) products whose risks are presented less diligently. On the other hand, they can rely on users having considerable experience with related processes, as well as a repertoire of cognitive and physical skills acquired in a wide variety of situations. Indeed, new product introductions can be particularly complicated when the target audience lacks relevant experience. Introductions may be quicker in the short run, but more expensive in the long run, when that audience puts too much faith in its existing knowledge and skills.

SOURCES FOR UNDERSTANDING THE PUBLIC

Just as it is sensible for laypeople confronted by a new product to look for familiar experiences and general knowledge, its producers might do the same when anticipating the response of that public. Responsible firms can be trusted to examine the specific experiences that members of the public have with their own and competitors' existing products. They may not, however, find their way to the general research literature on risk-related behavior. The next section of this paper is intended to improve access to that literature by summarizing general patterns and providing representative references. It draws primarily on the research literature in human judgment and decision making and its subspecialty focused on the perceptions of technological hazards. These fields are, roughly, 35 and 20 years old, respectively. Their roots are in the literatures on attitude change, clinical judgment, and human factors, each of which received a major push as part of the U.S. effort during World War II, as well as the much older fields of experimental psychology and decision theory.

These literatures provide substantive results that can be tentatively extrapolated to predict or explain people's responses to new products. For example, many studies have found that people are relatively insensitive to the extent of their own knowledge (Fischhoff et al., 1977; Wallsten and Budescu, 1983). The most common result is overconfidence, for example, being correct on only 80 percent of those occasions when one is absolutely certain of being correct. The generality of these findings (in those settings that have been studied) suggests that people might have undue confidence in their beliefs about new products and about how the attendant risks can be controlled. If this seems like a reasonable and worrisome hypothesis, then the research literature might be consulted for procedures able to improve people's judgment. For example, telling people that overconfidence is common seems to have little effect, whereas presenting people with personalized feedback regarding the appropriateness of their own confidence can make a positive difference (Fischhoff, 1982).

If one wanted to test these or other hypotheses, then the existing research also provides well-understood methodologies for conducting studies specific to particular risks. Ascertaining people's beliefs and values is a craft having as many nuances as does assessing their physiological functions or conducting measurements in the natural or biological sciences. For example, two formally equivalent ways of asking people to estimate how large a risk is can produce estimates that vary by several orders of magnitude (Fischhoff and MacGregor, 1983). A study that used one method might make it seem as though people underestimate the risk, whereas a study using the other method would produce apparent overestimates. As in other sciences, such measurement artifacts are sometimes predicted on the basis of general theories (Poulton, 1968, 1982), whereas in other cases they are discovered by trial and error. Exploiting this experience offers the opportunity to avoid the mistaken interpretations, and perhaps even mistaken policies, that such experimental artifacts can produce.

The applications of these methods to people's perceptions of technological hazards have seldom produced results challenging the overall conclusions from the general literature on judgment and decision making. These studies have, however, provided important elaborations, for example, showing just what people believe about particular risks, just how confident they are in those beliefs, or just how far they trust risk information coming from particular sources. They have also drawn attention to general issues with particular significance for consumer products and workplace processes, such as how people evaluate the trustworthiness of risk information (Baum et al., 1983; Johnson and Tversky, 1983; Richardson et al., 1987; Weinstein, 1987).

These detailed, systematic empirical studies stand in stark contrast to the casual observations that dominate many discussions of the public's behavior. Perhaps surprisingly, even scientists, who would hesitate to make any statements about topics within their own areas of competence without a firm research base, are willing to make strong statements about the public on the basis of anecdotal evidence. Unfortunately, immediate appearances can be deceiving, as when salient examples of public behavior are not particularly representative. And, as mentioned, even systematic observations can be misleading if not undertaken with a full understanding of the relevant methodology. An unfounded belief in having understood the public is a serious barrier to acquiring a genuine understanding.

The limits to casual observation might be seen in the coexistence of conflicting claims about the public, often associated with conflicting recommendations regarding how to deal with it. For example, advocates of deregulation frequently describe members of the public as understanding risks so well that they can readily fend for themselves in an unfettered marketplace. This confidence in the public is usually shared by those who advocate extensive public participation in risk management, through such avenues as hearings and information campaigns (Magat and Viscusi, 1992). Quite the opposite conclusion about public competence underlies proposals to leave risk management to technical experts or to force people to adopt risk-management practices that are "for their own good." Examples here include seatbelts, crash helmets, and dietary restrictions. Given the political and safety implications of these conflicting perceptions about the public, laypeople's behavior would seem to merit careful study. Good, hard evidence could provide guidance for managing risks, resolving conflicts between the public and technical experts, supplying the information that the public needs for better understanding, and creating technologies whose risks are acceptable to the public (Viscusi, 1992). The following section provides a summary of conclusions that can be drawn from studies of risk perception, as well as from the general research literature regarding judgment and decision making.

WHAT IS KNOWN

People Simplify

Most substantive decisions require people to deal with more nuances and details than they can readily handle at any one time. People have to juggle a multitude of facts and values when deciding, for example, whether to change jobs, trust merchants, or protest a toxic landfill. To cope with the overload, people simplify. Rather than attempting to think their way through to comprehensive, analytical solutions to decision-making problems, people try to rely on habit, tradition, the advice of neighbors or the media, and on general rules of thumb, such as nothing ventured, nothing gained. Rather than consider the extent to which human behavior varies from situation to situation, people describe other people as encompassing personality traits, such as being honest, happy, or risk seeking (Nisbett and Ross, 1980). Rather than think precisely about the probabilities of future events, people rely on vague quantifiers, such as "likely" or "not worth worrying about"—terms that are used differently in different contexts and by different people (Beyth-Marom, 1982).

The same desire for simplicity can be observed when people press risk managers to categorize technologies, foods, or drugs as "safe" or "unsafe," rather than to treat safety as a continuous variable. It can be seen when people demand convincing proof from scientists who can provide only tentative findings. It can be seen when people attempt to divide the participants in risk disputes into good guys and bad guys, rather than viewing them as people who, like themselves, have complex and interacting motives. Although such simplifications help people to cope with life's complexities, they can also obscure the fact that most risk decisions involve gambling with people's health, safety, and economic well-being in arenas with diverse actors and shifting alliances.

Once People's Minds Are Made Up, It Is Hard to Change Them

People are quite adept at maintaining faith in their current beliefs unless confronted with concentrated and overwhelming evidence to the contrary. Although it is tempting to attribute this steadfastness to pure stubbornness, psychological research suggests that some more complex and benign processes are at work (Nisbett and Ross, 1980).

One psychological process that helps people to maintain their current beliefs is feeling little need to look actively for contrary evidence. Why look if one does not expect that evidence to be very substantial or persuasive? For example, how many environmentalists read the *Wall Street Journal* and how many industrialists read the Sierra Club's *Bulletin* to learn something about risks (as opposed to reading these publications to anticipate the tactics of the opposing side)? A second contributing thought process is the tendency to exploit the uncertainty surrounding apparently contradictory information in order to interpret it as being consistent with existing beliefs (Gilovich, 1993). In risk debates, a stylized expression of this proficiency is finding just enough problems with contrary evidence to reject that evidence as inconclusive.

A third thought process that contributes to maintaining current beliefs can be found in people's reluctance to recognize when information is ambiguous. For example, the incident at Three Mile Island would have strengthened the resolve of any antinuclear activist who asked only, "How likely is such an accident, given a fundamentally unsafe technology? —just as it would have strengthened the resolve of any pronuclear activist who asked only, "How likely is the containment of such an incident, given a fundamentally safe technology?" Although a very significant event, Three Mile Island may not have revealed very much about the riskiness of nuclear technology as a whole. Nonetheless, it helped the opposing sides to polarize their views. Similar polarization followed the accident at Chernobyl, with opponents pointing to the consequences of a nuclear accident, which they see as coming with any commitment to nuclear power, and proponents pointing to the unique features of that particular accident, which are unlikely to be repeated elsewhere, especially considering the precautions instituted in its wake (Krohn and Weingart, 1987).

People Remember What They See

Fortunately, given their need to simplify, people are good at observing those events that come to their attention and that they are motivated to understand (Hasher and Zacks, 1984; Peterson and Beach, 1967). As a result, if the appropriate facts reach people in a responsible and comprehensible form before their minds are made up, there is a decent chance that their first impression will be the correct one. For example, most people's primary sources of information about risks are what they see in the news media and observe in their everyday lives. Consequently, people's estimates of the principal causes of death are strongly related to the number of people they know who have suffered those misfortunes and the amount of media coverage devoted to them (Lichtenstein et al., 1978).

Unfortunately, it is impossible for most people to gain first-hand knowledge of many hazardous technologies. Rather, what laypeople see are the outward manifestations of the risk-management process, such as hearings before regulatory bodies or statements by scientists to the news media. In many cases, these outward signs are not very reassuring. Often, they reveal acrimonious disputes between supposedly reputable experts, accusations that scientific findings have been distorted to suit their sponsors, and confident assertions that are disproven by subsequent research (MacLean, 1987; Rothman and Lichter, 1987).

Although unattractive, these aspects of the risk-management process can provide the public with potentially useful clues to how well technologies are understood and managed by industry and regulatory agencies. Presumably, people evaluate these clues just as they evaluate the conflicting claims of advertisers and politicians. It should not be surprising, therefore, that the public sometimes comes to conclusions that differ from what risk managers hope or expect. For example, it was reasonable to conclude that saccharin is an extremely potent carcinogen after seeing the enormous scientific attention that it generated some years back. Yet, much of the controversy actually concerned how to deal with a food that was strongly suspected of being a weak carcinogen. In some cases, the public may have a better overview on the proceedings than the scientists and risk managers mired in them, realizing perhaps that neither side knows as much as it claims.

People Cannot Readily Detect Omissions in the Evidence They Receive

Unfortunately, not all problems with information about risk are as readily observable as blatant lies or unreasonable scientific hubris. Often the information that reaches the public is true, but only part of the truth. Detecting such systematic omissions proves to be difficult (Tversky and Kahneman, 1973). For example, most young people know relatively few people suffering from the diseases of old age, nor are they likely to see those maladies cited as the cause of death in newspaper obituaries. As a result, young people tend to underestimate the frequency of these causes of death, while most people overestimate the frequency of vividly reported causes, such as murder, accidents, and tornadoes (Lichtenstein et al., 1978).

Laypeople are even more vulnerable when they have no way of knowing about information that has not been disseminated. In principle, for example, one could always ask physicians if they have neglected to mention any side effects of the drugs they prescribe. Likewise, people could ask merchants whether there are any special precautions for using a new power tool, just as they could ask proponents of a hazardous facility if their risk assessments have considered all forms of operator error and sabotage. In practice, however, these questions about omissions are rarely asked. It takes an unusual turn of mind and personal presence to recognize one's own ignorance and insist that it be addressed.

As a result of this insensitivity to omissions, people's risk perceptions can be manipulated in the short run by selective presentation. Not only will people not know what they have not been told, but they will not even feel how much has been left out (Fischhoff et al., 1978). What happens in the long run depends on whether the unmentioned risks are revealed by experience or by other sources of information. When deliberate omissions are detected, the responsible party is likely to lose all credibility. Once a shadow of doubt has fallen, it is hard to erase.

People May Disagree More about What Risk Is Than about How Large It Is

Given this mixture of strengths and weaknesses in the psychological processes that generate people's risk perceptions, there is no simple answer to the question, How much do people know and understand? The answer depends on the risks and on the opportunities that people have to learn about them.

One obstacle to determining what people know about specific risks is disagreement about the definition of "risk" (Crouch and Wilson, 1981; Fischhoff et al., 1983; Fischhoff et al., 1984; Slovic et al., 1979). The opportunities for disagreement can be seen in the varied definitions used by different risk managers. For some, the natural unit of risk is an increase in probability of death; for others, it is reduced life expectancy; for still others, it is the probability of death per unit of exposure, where "exposure" itself may be variously defined.

The choice of definition is often arbitrary, reflecting the way in which a particular group of risk managers habitually collects and analyzes data. The choice, however, is never trivial. Each definition of risk makes a distinct political statement regarding what society should value when it judges the acceptability of risks. For example, "reduced life expectancy" puts a premium on deaths among the young, which would be absent in a measure that simply counted the expected number of premature deaths. A measure of risk could also give special weight to individuals who can make a special contribution to society, to individuals who were not consulted (or even born) when a risk-management policy was enacted, or to individuals who do not benefit from the technology generating the risk.

If laypeople and risk managers use the term "risk" differently, then they can agree on the facts about a specific technology but still disagree about its degree of riskiness. Some years ago, the idea circulated in the nuclear power industry that the public cared much more about multiple deaths from large accidents than about equivalent numbers of casualties resulting from a series of small accidents. If this assumption were valid, the industry would be strongly motivated to remove the threat of such large accidents. If removing the threat proved impossible, then the industry could argue that a death is a death and that, in formulating social policy, it is totals that matter, not whether deaths occur singly or collectively.

There were never any empirical studies to determine whether this was really how the public defined risk. Subsequent studies, though, have suggested that what bothers people about catastrophic accidents is the perception that a technology capable of producing such accidents cannot be very well understood or controlled (Slovic et al., 1984). From an ethical point of view, worrying about the uncertainties surrounding a new and complex technology, such as nuclear power, is different from caring about whether a fixed number of lives is lost in one large accident rather than in many small accidents.

People Have Difficulty Detecting Inconsistencies in Risk Disputes

Despite their frequent intensity, risk debates are typically conducted at a distance (Krimsky and Plough, 1988; Mazur, 1973; Nelkin, 1978). The disputing parties operate within self-contained communities and talk principally to one another. Opponents are seen primarily through their writing or their posturing at public events. Thus, there is little opportunity for the sort of subtle probing needed to discover basic differences in how the protagonists think about important issues, such as the meaning of key terms or the credibility of expert testimony. As a result, it is easy to misdiagnose one another's beliefs and concerns.

The opportunities for misunderstanding increase when the circumstances of the debate restrict candor. For example, some critics of nuclear power actually believe that the technology can be operated with reasonable safety. However, they oppose it because they believe that its costs and benefits are distributed inequitably. Although they might like to discuss these issues, public hearings about risk and safety often provide these critics with their only forum for venting their concern. If they oppose the technology, then they are forced to do so on safety grounds, even if this means misrepresenting their perceptions of the actual risk. Although this may be a reasonable strategy for pursuing their ultimate goals, it makes them look unreasonable to observers who hold opposing views of nuclear power.

Individuals also have difficulty detecting inconsistencies in their own beliefs or realizing how simple reformulations would change their perspectives on issues. For example, most people would prefer a gamble with a 25 percent chance of losing $200 (and a 75 percent chance of losing nothing) to a sure loss of $50. However, most of the same people would also buy a $50 insurance policy to protect against such a loss. What they will do depends on whether the $50 is described as a "sure loss" or as an "insurance premium." In such cases, one cannot predict how people will respond to an issue without knowing how they will perceive it, which depends, in turn, on how it will be presented to them by merchandisers, politicians, or the media (Fischhoff, 1991; Fischhoff et al., 1980; Turner and Martin, 1984; Tversky and Kahneman, 1981).

Thus, people's insensitivity to the nuances of how risk issues are presented exposes them to manipulation. For example, a risk might seem much worse when described in relative terms, such as doubling their risk, than in absolute terms, as in increasing that risk from one in a million to one in a half million. Although both representations of the risk might be honest, their impacts would be quite different. Perhaps the only fair approach is to present the risk from both perspectives, letting recipients determine which one, or hybrid, best represents their world view.

Experts Are People, Too

Obviously, experts have more substantive knowledge than laypeople. Often, however, the practical demands of risk management force experts to make educated guesses about critical facts, taking them far beyond the limits of their data. In such situations, debates about risk are often conflicts between competing sets of risk perceptions, those of the public and those of the experts. As a result, one must ask how good those expert judgments are. Do experts, like laypeople, tend to exaggerate the extent of their own knowledge? Are experts more sensitive than others to systematic omissions in the evidence that they receive? Do they, too, tend to oversimplify policy issues?

Available studies suggest that when experts must rely on judgment, their thought processes often resemble those of laypeople (Fischhoff, 1989; Kahneman et al., 1982; Mahoney, 1979; Shlyakhter et al., 1994). For example, Figure 1 displays two cases of overconfidence in the judgments of se-

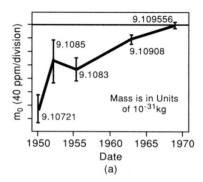

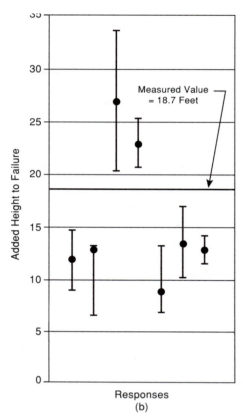

FIGURE 1 Two examples of overconfidence in expert judgment. Overconfidence is represented by the failure of error bars to contain the true value. (a) Subjective estimates by senior scientists of the rest mass of the electron. (b) Subjective estimates by civil engineers of the height at which a test embankment would fail. In each example, the range of values provided by the experts failed to contain the true value as it was later determined. SOURCE: Fischhoff and Svenson (1988).

nior scientists in their attempts to determine ranges of possible values (or confidence intervals) for topics within their areas of expertise (Henrion and Fischhoff, 1986; Hynes and Vanmarcke, 1976). Anecdotal evidence of the limits to expert judgment can be found in many cases where fail-safe systems have gone seriously awry or where confidently advanced theories have been proved wrong. For example, DDT came into widespread and uncontrolled use before the scientific community had seriously considered the possibility of side effects. Medical procedures, such as using DES to prevent miscarriages, sometimes produce unpleasant surprises after being judged safe enough to be used widely (Berendes and Lee, 1993; Grimes, 1993). The accident sequences at Three Mile Island and Chernobyl seem to have been out of the range of possibility, given the theories of human behavior underlying the design of those reactors (Aftermath of Chernobyl, 1986; Hohenemser et al., 1986; Reason, 1990). Of course, science progresses by absorbing lessons that prompt it to discard incorrect theories. However, society cannot make rational decisions about hazardous technologies without knowing how much confidence to place in them and in the scientific theories on which they are based.

In the academic community, vigorous peer review offers technical experts some institutional protection against the limitations of their own judgments. Unfortunately, the exigencies of risk management, including time pressures and resource constraints, often strip away these protections, making risk assessment something of a quasi science (like much cost/benefit analysis, opinion polling, or evaluation research), bearing more of the rights than the responsibilities of a proper discipline.

On the basis of psychological theory, one would trust experts' opinions most where they have had the conditions needed to learn how to make good judgments. These conditions include prompt, unambiguous feedback that rewards them for candid judgment and not, for example, for exuding confidence. Weather forecasters do have these conditions, and the result is a remarkable ability to assess the extent of their own knowledge (Murphy and Winkler, 1984). It rains almost exactly 70 percent of the time when they are 70 percent confident that it will. Unfortunately, such conditions are rare. When feedback is delayed, as with predictions of the carcinogenicity of chemicals having long latencies, learning may be very difficult. Further problems arise when expert predictions are ambiguous or the lessons of subsequent experience are hard to unravel. Psychological theory also suggests that learning is likely to be fairly local. Thus, one might question toxicologists' judgments about social policy just as much as social policymakers' judgments about toxicology (Cranor, 1993).

APPLYING BEHAVIORAL RESEARCH

The Interface of Products, People, and Law

Manufacturers have a legal obligation to produce products that are "duly safe" (Wade, 1973). Unfortunately, the law fails to specify clearly how much safety is required. One source of uncertainty is the lack of detailed feedback from the courts (Saks, 1992), making it difficult for citizens or scholars to discern general patterns (Eisenberg and Henderson, 1993; Henderson, 1991; Merz, 1991a). The news media may compound problems by disproportionately reporting cases with unusual fact situations or particularly large punitive damage awards. Much less attention is devoted to the remitter of that award or subsequent settlement or reversal on appeal (Nelkin, 1984; Rustad, 1992).

Even if more detailed information were available regarding litigated cases, that information might be of limited usefulness because of systematic biases in the choices of cases for trial. Lawsuits arise from a very small proportion of claims that are, in turn, a small subset of the injuries associated with a product (Hensler et al., 1991; Localio et al., 1991). Thus, manufacturers receive meager signals from the courts regarding how to manage their affairs (Huber and Litan, 1991). Perhaps the best they can do to anticipate legal problems is compare the attributes of their product with those of existing products. When they share features with products that have proven problematic, they should make extraordinary efforts at each stage of the design, testing, production, marketing, and post-marketing process (Weinstein et al., 1978).

Product liability may be viewed as attempting to regulate these various aspects of a manufacturer's design, production, and marketing behaviors. To explore this aspect of the law, it is useful to draw an analogy to the FDA licensing process as a prototype for product development, production, and marketing, because this process is the one our society uses to ensure that new drugs are acceptably safe.[1] Figure 2 conceptualizes that process to highlight the interaction between product liability law and manufacturers' product decisions. This figure presents a discrete set of steps and a key decision. All of the steps must be taken adequately and the decision must be answered correctly in order to shield a manufacturer from liability for any injuries caused by the resulting products. We depict the process as a loop, because the obligation to collect information and to act on it transcends any one sale, leading to a continuous process of learning (e.g., how the product is actually used or misused by consumers) and modification of the marketing process.[2] Liability may result from departures from the "expected" norms for the management of product-related information (e.g., regarding the risks and benefits of the products) and

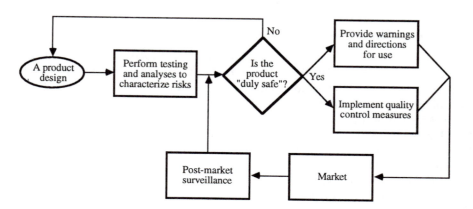

FIGURE 2   Simplified view of a generalized product marketing processs based on the Food and Drug Administration licensing process.

decision making. Absent regulatory oversight, we do not propose that following the process in itself is a defense, but, rather, that following this process may assist manufacturers in placing into the stream of commerce products that are "duly safe."

In detail, the steps in this process are as follows:

1. *Testing and analysis to characterize risks*—performing reasonable research and testing to ascertain the risks and benefits of alternative designs arising from foreseeable uses of the product, considering (a) who the target consumers and users of the products are and whether these users are particularly susceptible to injury from the product, (b) the potential for misuse and abuse of the product, (c) the nature of the injuries caused and the base rate of similar injuries in the user population attributable to similar products or nonattributable to any known causes, (d) the state of scientific knowledge about the risks, and (e) uncertainties in the above factors.

2. *Determining whether the product is duly safe*—making design choices to ensure that the products are "duly safe" by balancing the risks and benefits of alternative designs, expressly considering (a) the state of the art, (b) the reasonable availability and practicability of modifications and safety enhancements, (c) regulatory and industrial standards applicable to product safety, and (d) features of competing products.

3. *Providing warnings and directions for use*—providing warnings of any remaining latent and nonobvious risks so that consumers can make informed choices about product purchase and use and to provide adequate directions to enable them to use the products properly, taking into account (a) the obviousness of the risks, (b) whether a warning is feasible (i.e., whether users are children, the risk is to a third party, or there is no practicable way to place a useful warning on the product), and (c) labeling standards, if any.

4. *Quality control measures*—implementing adequate production techniques to assure that products meet design specifications, and instituting quality control mechanisms to reduce the chance of release of any products with undetected flaws, taking into account the probability and nature of any injuries that may be associated with such flaws and the marginal costs of reducing those risks.

5. *Marketing*—marketing the product in a responsible manner, ensuring that advertising and promotion techniques do not constitute warranties, misrepresent the product, or target especially susceptible users for whom the product design, instructions, or warnings are inadequate.

6. *Post-market surveillance*—monitoring product performance in the marketplace and modifying product design, production, and warnings,

recalling products for correction, or pulling the product off the market in response to user feedback (Lamken, 1989).

Many of the decisions that manufacturers must make in the process outlined above are extremely difficult. To help inform these decisions, we first link the process with product liability law, then identify what the law requires at each step of the process. Product liability encompasses several distinct doctrines, each providing an independent basis on which liability can be predicated. Each doctrine places different obligations on a manufacturer and seller. Each offers different opportunities for using an understanding of how consumers perceive products in order to reduce their product liability exposure. The first doctrine is that the way a product is represented to consumers may give rise to warranties, whether intended or unintended. Painting an unduly glowing picture of a product may increase sales. However, it may also mislead consumers into thinking that there are differential quality and safety features, which could, in turn, lead reasonable consumers to lower their guard inappropriately. The law obliges the seller to ensure that seller and buyer have the same product in mind at the time of a sales transaction.

Second, manufacturers will be liable for injuries caused by a "defective" product (American Law Institute, 1965). A product can be deemed defective if it has a manufacturing defect or flaw, if its design is defective, or if the warnings and instructions provided with the product are defective. Thus, an injured consumer has several tiers of alleged defectiveness on which to base a claim for damages, and manufacturers must run a gauntlet to avoid liability. Figure 3 depicts the three bases of a plaintiff's claim for injuries caused by an allegedly defective product, each of which provides an independent and sufficient basis for liability.

At the most basic level, sellers will be liable only for injuries caused by

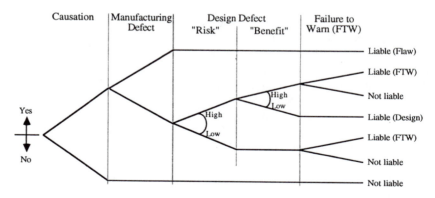

FIGURE 3   A logically structured schema of product liability law.

their products. The causal inquiry may be extremely complex because of uncertain exposures, confounding causes and base rates of injury in the population at risk, and limited scientific knowledge about causal relationships (Brennan, 1988). In the figure, causation is presented deterministically to reflect the strong dependence of liability on it.

Moving across the figure to the right, the diagram depicts strict liability for manufacturing defects or flaws. Manufacturers are held strictly liable for products that deviate from the intended design or from expected or normal production (Henderson, 1973). The producer must make a trade-off between the real costs of the quality controls needed to avoid such manufacturing flaws and the potential costs of injury and liability from substandard products.

The third basis of liability is design defect, meaning either that the benefits of the product did not justify the risks or that the design failed to meet consumers' reasonable expectations for that product (Henderson, 1979). In either case, the manufacturer must decide how safe a product must be to justify its benefits as well as the utility that would be lost were the product made safer. Liability will usually turn on the availability and feasibility of alternative product designs, which could have made the product safer with no or little loss of consumer utility. Understanding how people use different products and perceive risks and benefits is a key to making good design decisions (Chapanis, 1959). It is also important for a manufacturer to be aware of technological developments as they define the state of the art, to keep abreast of the design standards set by competitors, and to comply with industry and regulatory requirements. Deviations from expected norms, as embodied in industry practices and applicable standards, may provide strong evidence of substandard design. Justifications for any such deviations must be embodied explicitly in analysis, testing, and managerial decision making (Twerski et al., 1980).

The fourth basis for product liability arises from inadequate directions for the proper use of a product or from failure to provide warnings of known but nonobvious risks associated with its use (Henderson and Twerski, 1990). Regardless of the utility of the low-risk product, any known nonobvious or latent risk must be disclosed to avoid potential liability. This obligation may extend to providing subsequently acquired information to consumers after the sale, or even to product redesign, recall, and modification. The duty to disclose is not absolute, however, because consumers are generally charged with knowing open and notorious hazards, although exactly what such hazards are may be a question of interpretation. Here, too, better understanding of consumers' perceptions, as well as their interpretations of warnings and instructions when provided, will lead to better and more defensible products.

The following sections discuss several general strategies for exploiting the research described above in order to meet these conditions. Each successive option affords an increasingly central role to concern about the public in the product-stewardship process. The analysis begins with the reactive approach of attempting to predict the extent of problems arising from the way a product is used and ends with the proactive approach of basing product design on concern about risk to users.

Predicting Product-Use Problems

The least that manufacturers can do is to face up to their problems. A systematic analysis of the likelihood and severity of incidents is essential to any reasoned allocation of resources in product management. It can help determine how much to spend on different kinds of design, testing, training, and warning. It may even identify cases where a product is too risky to manufacture, considering the direct costs of compensating those who are injured and the indirect costs of diminished reputation. Such analyses must be performed during the initial design and testing of a product and continue through product marketing and post-market surveillance. The goal is to ensure that products are marketed to consumers without creating unwarranted and unwanted representations of the quality and functionality of the product.

As many of the other chapters in this volume indicate, many manufacturers have procedures in place for making such determinations. Behavioral research might contribute to these processes in several ways. One is identifying circumstances that call for analysis. For example, many products are in constant development, as manufacturers generate new styles or exploit advances in materials or fabrication techniques. In addition to these changes over time, many products are simultaneously manufactured in a variety of related forms to suit the needs of different market niches.

A careful analysis might reveal surprises in how these alternative versions are used. At times, even cosmetic changes can affect the way a product is used, such as by suggesting less need for caution or a broader range of applications, not all of which were intended. At other times, meaningful changes may go unnoticed, leading to negative transfer—the interference of old habits with new situations.[3] At still other times, users may simply need time to master the quirks of the new version of a familiar product. Until they do, they are exposed to "predictable surprises."

Ironically, these transitional problems can be exacerbated by the natural desire for design improvement. A system may be in constant flux, so that no configuration is ever really mastered. Its rate of problems and near problems may be too low to allow for learning. Training exercises may be unrealistic, or telegraph their punches, or evoke abnormal levels of vigilance. Automatic controls, such as aircraft landing systems, may "deskill" users, depriving them of a feeling for how the system operates and decreasing their ability to respond when human intervention is needed. Of course, many products suffer from persistent problems rather than frustrating near-perfection (National Research Council, 1993).

Assuming that problems seem possible, one must estimate their probability and severity, as critical inputs to deciding whether to pursue a project or invest in redesign. With a stable system, one can learn a great deal from properly conducted statistical analyses. Those estimates become less reliable when things (the product, the users, the uses, the usage environment) change. In such cases, the impact of those changes must be assessed.

The most elaborate assessments are applications of probabilistic risk assessment (PRA) to estimate the reliability of systems with human components. These procedures attempt to reduce complex, novel systems to components whose reliability can be estimated on the basis of historical data. Although these analyses are standard practice in many settings, they must be interpreted cautiously, especially when used to estimate the magnitude of risks and not just to clarify the relationships among system components. Even if ample historical data are available at the component level, identifying the relevant components and relationships requires the exercise of judgment. Further judgment is needed to adapt general data to specific circumstances, to assess the uncertainty in the estimates, and to determine the definitiveness of the analysis. A good deal is known about the imperfections in these processes and the procedures for making the best use of expert judgment. They build, in part, on research showing the judgmental problems encountered by technical experts when they run out of data and rely on educated intuition. Unfortunately, not all analyses follow the best procedures (Morgan and Henrion, 1991).

Done right, a good PRA can be much better than no analysis at all. Done uncritically, it can create an illusion of understanding. Done with the aim of demonstrating (rather than estimating or improving) safety, it can be a tool of deception and self-deception. Even the best and best-understood estimates will be useful only if there is a legal and logical decision-making scheme for using them. They may be worthless if management cannot face an unpleasant message. They may be injurious if a firm is penalized for thinking explicitly about risk issues, as in some interpretations of the Ford Pinto case.

Warning Users about Potential Risks

The least that one can do when a product poses nonnegligible risks is to inform possible users, allowing them to determine whether the benefits justify those risks. In some cases, doing so will allow producers to claim that users have provided informed consent, assuming upon themselves responsibility for any adverse consequences.

Selection. The first step in designing warnings is to select the information that they should contain. In many communications, that choice seems fairly arbitrary, reflecting some technical expert or communication specialist's notion of "what people ought to know." Poorly chosen information can have several negative consequences, including both wasting recipients' time and being seen to waste it. It exposes recipients to being judged unduly harshly if they are uninterested in information that, to them, seems irrelevant. In its fine and important report, *Confronting AIDS*, the Institute of Medicine (1986) despaired after a survey showed that only 41 percent of the U.S. public knew that AIDS was caused by a virus. Yet, one might ask what role that information could play in any practical decision, as well as what those people who answered correctly meant by "a virus."

One logically defensible criterion is to use value-of-information analysis to identify those messages that ought to have the greatest impact on pending decisions. Such information would resolve major uncertainties for recipients regarding the probability of incurring significant outcomes, as the result of taking different actions. Doing so requires taking seriously the details of potential users' lives, looking hard at the options they face and the goals they seek.

Merz (1991b; Merz et al., 1993) applied value-of-information analysis to a well-specified medical decision, whether to undergo carotid endarterectomy. Both this procedure, which involves scraping out an artery that leads to the head, and its alternatives have a variety of possible positive and negative effects. These effects have been the topic of extensive research, which has provided quantitative risk estimates of varying precision. Merz created a population of hypothetical patients, who varied in their physical condition and relative preferences for different health states. He found that knowing about a few, but only a few, of the possible side effects would change the preferred decision for a significant portion of patients. He argued that communications focused on these few side effects would make better use of patients' attention than laundry lists of undifferentiated possibilities. He also argued that his procedure could provide an objective criterion for identifying the information that must be transmitted to ensure that patients were giving a truly informed consent.

Although laborious, such analyses offer a possibility of closure that is unlikely with the traditional ad hoc procedures of relying on professional judgment or conventional practice.

Presentation. Once information has been selected, it must be presented in a comprehensible way. Many of the principles of presentation are well studied and established. For example, research has shown that comprehension improves when text has a clear structure and, especially, when that structure conforms to recipients' intuitive representations; that critical information is more likely to be remembered when it appears at the highest level of a clear hierarchy; and that readers benefit from "adjunct aids," such as highlighting, previews (showing what to expect), and summaries. Such aids can even be better than full text for understanding, retaining, and being able to look up information (e.g., Ericsson, 1988; Garnham, 1987; Kintsch, 1986; Reder, 1985; Schriver, 1989).

As suggested by the research reviewed earlier, information about the magnitude of risks poses some particular challenges to communicators. The units, orders of magnitude, and even the very idea of quantitative risk estimates may be foreign to many recipients. Under those circumstances, it may seem appealing to provide no more than a general indication of risk levels. However, that situation may also produce the greatest variability in the magnitudes attributed to such verbal quantifiers. The very fact that risks are mentioned may suggest that they are relatively severe, even though that act reflects no more than the caution of a particular producer or the idiosyncrasies of a particular legislative or regulatory process (which mandated labeling).

Although there may be no substitute for providing explicit quantitative information, there is also no guaranteed way to do it effectively. For the time being, we must resign ourselves to an imperfect process in which producers gradually learn how to communicate and users gradually learn how to understand. Fortunately, many decisions are relatively insensitive to the precision of perceived risk estimates (von Winterfeldt and Edwards, 1986). As a result, imperfect communication may still allow people to identify the courses of action in their own best interests. Recipients can always choose to ignore quantitative information or to convert it to some intuitive qualitative equivalent (Beyth-Marom, 1982; Wallsten et al., 1986). In the domain of weather forecasting, studies have found that people appreciate the quantitative information in probability of precipitation forecasts, although they are sometimes unsure about exactly what event is being forecast (Krzysztofowicz, 1983; Murphy and Brown, 1983; Murphy et al., 1980).

Intuitively recognizing the difficulty that people may have with understanding small risks, technical experts have often sought to provide perspective by embedding a focal risk in a list of equivalent risks. Those lists might show the doses of a variety of activities that produce one chance in a million of premature death, for example, hours of canoeing, teaspoonfuls of peanut butter, years living near the boundary of a nuclear power plant, or the loss of life expectancy from various states and activities (Cohen and Lee, 1979; Wilson, 1979). Assuming that recipients had accurate feelings for the likelihood of some of the items in the list, this strategy might be useful. Unfortunately, risk comparisons are often formulated with transparently rhetorical purpose, attempting to encourage recipients' acceptance of the focal risk—"if you like peanut butter, and accept its risks from aflatoxin, then you should love nuclear power." By failing to consider the other factors entering into others' decisions, for example, the respective benefits from peanut butter and nuclear power, such comparisons have no logical force (Fischhoff et al., 1981; Fischhoff et al., 1984). They can, however, alienate recipients, who prefer to make their own decisions. To date, there is no clear demonstration of risk comparisons being an effective communication technique (Covello et al., 1988; Roth et al., 1990).

Improving the Usage of Existing Products

Telling people about residual risks is, in effect, an admission of engineering failure. It says, "this is the best that we could do, see if you want to live with it." To some extent, that failure may be inevitable. Engineers cannot do it all, and must rely on responsible use of their products. Indeed, in many situations, the product user is in the best position to assess and minimize the risks. However, to fulfill their obligation, engineers must provide users with the information that they need for successful operation. Providing estimates of the magnitude of potential risks (discussed in the previous section) is a part of that story, insofar as it helps users decide how seriously to take safety issues.

Additional steps range from general warnings ("poisonous"), to specific warnings ("use in a well-ventilated area"), to detailed instructions, to training courses of various length. The challenge in designing such materials and procedures is to achieve an acceptable level of understanding at minimal cost in time, money, and effort to producer and user. The required understanding might be defined as whatever is needed to achieve that announced safety level, assuming responsible use.

Training is a heavily studied topic, another beneficiary of the demands of World War II and the Cold War. A producer who did not exploit its potential would arguably bear some responsibility for whatever misfortunes followed. However, it could not be expected to eliminate all risks. Many modern products are so complex and used in such diverse circumstances that it is impossible to anticipate all contingencies or to train users to the desired proficiency. One response to this reality has been to shift the focus of training from what to do to what is happening. It attempts to provide users with accurate mental models of how the product works, so that they can anticipate the results of their actions and diagnose potentially problematic circumstances. Doing so requires a user-centered, rather than a product-centered approach. It might be advantageous also in situations that less obviously strain the limits of traditional training (Laughery, 1993; Norman, 1988; Reason, 1990).

The state of the art for written instructions seems rather more primitive. As with informing users about the magnitude of risks, the selection and organization of operational information often seem somewhat arbitrary. To take an example that we have studied intensively, the U.S. Environmental Protection Agency (EPA) invested heavily in the development, evaluation, and dissemination of a *Citizen's Guide to Radon*. Yet, despite this laudably broad and deep effort, the resulting brochure uses a question-and-answer format, with little attempt to summarize or impose a general structure. It relies on an untested risk-comparison scale, which, for example, uses relatively similar risks, but presents them on a logarithmic scale. It does not explicitly confront a misconception that could undermine the value of other correct beliefs about radon: because radon is radioactive, it can contaminate permanently, making it infeasible to remediate should a problem be found (hence not even worth testing).

That misconception was identified in a series of open-ended interviews intended to characterize laypeople's mental models of this risk and confirmed in studies using structured questionnaires (Bostrom, 1990; Bostrom et al., 1992). This set of procedures was used to examine lay understanding of a variety of risks, including those from electromagnetic fields, Lyme disease, lead poisoning, and nuclear energy sources in space (Fischhoff et al., 1993; Morgan et al., 1992). Leventhal and his colleagues have used similar approaches in studying adherence to medical regimes such as routine screening, hypertension drugs, and diets (Leventhal and Cameron, 1987). Yet other investigators have looked at lay conceptualizations of such diverse processes as macroeconomics, physics, computers, and climate change (Carroll and Olson, 1988; Chi et al., 1981; Jungermann et al., 1988; Kempton, 1991; MacGregor, 1989; Voss et al., 1983). Typically, these studies find a mixture of accurate beliefs, on which instructions can build; misconceptions, which need to be eliminated; peripheral beliefs, which need to be placed in proper perspective; and vague beliefs, which need to be sharpened before they can be used, or judged for accuracy. Such procedures provide one of the only ways of identifying beliefs that would not have occurred to technical experts.

Open-ended procedures also provide one of the few ways of identifying discrepancies in how terms are used by people from different linguistic communities. Consider, for example, the common, simple warning, "Don't drink and drive." Recipients could hear, but not get, the message if they guessed wrong about what was meant in terms of the kind and amount of "drinking" and "driving," not to mention any special pleading as far as how the general message applied to them personally (Svenson, 1981). Quadrel (1990) asked adolescents to estimate risks using deliberately vague questions, such as, "What is the probability of getting into an accident if you drink and drive?" She found that even teens with poor education were quite sensitive to imprecisions in how risks were described. There was also considerable variability in how her subjects "filled in the blanks," in the sense of supplying missing details. As a result, they ended up answering different questions even when looking at the same words. As mentioned, earlier studies found that the disagreements between experts and laypeople about the magnitude of risks are due, in part, to disagreements about the definition of "risk" (Slovic et al., 1979; Vlek and Stallen, 1980).

Effective risk communications can help people to reduce their health risks or to get greater benefits in return for those risks that they do take. Ineffective communications not only fail to do so but also incur opportunity costs, in the sense of occupying the place (in recipients' lives and society's functions) that could be taken up by more effective communications. Even worse, misdirected communications can prompt wrong decisions by omitting key information or failing to contradict misconceptions, create confusion by prompting inappropriate assumptions or by emphasizing irrelevant information, and provoke conflict by eroding recipients' faith in the communicator. By causing undue alarm or complacency, poor communications can have greater public health impact than the risks that they attempt to describe. It may be no more acceptable to release an untested communication than an untested drug. Because communicators' intuitions about recipients' risk perceptions cannot be trusted, there is no substitute for empirical validation (Fischhoff, 1987; Fischhoff et al., 1983; National Research Council, 1989; Rohrmann, 1990; Slovic, 1987). Failing to develop and test messages systematically raises legal, ethical, and management questions.

Improving Product Design

Instructing users in how to avoid potential problems leads, in effect, to teaching them how to make the best of an imperfect situation. A more satisfying response is designing user problems away. That means treating users as a resource, rather than as a source of difficulties. Understanding their problems might mean gaining market share, as well as avoiding litigation. Even an explicit commitment to safety and operability can mean something in the marketplace. Delivering on that commitment could be worth even more.

Industries, companies, and even divisions differ greatly in their commitment to the human factors engineering needed to achieve operability. For example, on the same plane, one might find fancy cockpit displays, clumsy tray tables, and metal coffee pots that induce carpal tunnel syndrome. Inattention to these issues may reflect disinterest in the users (pilots may matter more than flight attendants) or simply the pecking order in design departments. An examination of the skills appearing in a firm's organizational chart provides one indication of how seriously it takes human factors, and of how well it is positioned to exploit the opportunities.

A second indicator of a firm's ability to improve its design is its reluctance to attribute mishaps to human (or user or operator) error. The demands of product liability suits may force a firm to make and defend such claims. However, making improvements requires sharing responsibility. The sort of user-centered, or mental-models, procedure described earlier provides one place to start that process. It means trying to bring designs closer to users' expectations, rather than vice versa (as discussed in the previous section).

One common source of discouragement is sometimes called the theory of "risk homeostasis" (Wilde, 1982). It holds that users frustrate safety improvements by finding ways to use products more aggressively. The evidence supporting this hypothesis is mixed (Slovic and Fischhoff, 1983). Were it true, it might suggest that users are so irrational that there is no point in trying to improve safety. An alternative interpretation would be that users are responding rationally, trying to get more benefit from a product, at the price of forfeiting the increase in safety. If users understand the risks and benefits involved, then they have, arguably, given informed consent for whatever happens. Their desire for greater benefit suggests a design opportunity: providing that benefit without sacrificing safety.

CONCLUSION

The suggestions in the preceding sections dealt with strategies that are within the control of a product's manufacturer. They are, in effect, proposals for improved product stewardship. How effective each proposal is depends, in part, on how skillfully it is implemented and, in part, on how good it conceivably could be. The limits to performance depend, in turn, partly on the strategy. A warning label cannot do as much as a user-centered redesign, although it may be the most cost-effective response.

Those limits also depend on the environment and how it rewards or punishes different strategies. For example, there is no incentive to sweat the details of message design if presenting a laundry list of potential problems is construed as ensuring informed consent. There is a disincentive for doing so if changing how risks are described can be construed as admitting the inadequacy of previous descriptions, which may be in litigation. There may be a disincentive for creating safer designs if that, too, can be construed as an admission of previous failure, or if stricter demands are made of new products. Firms may be penalized for rigorous testing if they can later be accused of releasing products with imperfections that they themselves have documented.

In addition to removing such obstacles to considering behavioral issues, incentives are needed to take them seriously. Generally speaking, firms should get credit for vigorously studying the behavior of potential users, for having behavioral specialists involved throughout the design process, for evaluating the residual risks of their products, for communicating both the extent and the sources of those risks to users, and for measuring how successfully those messages have gotten across.

NOTES

1. The FDA licensing process is laid out in 21 Code of Federal Regulations, Parts 310, 312, and 314 (1993). Manufacturing requirements are set forth in Parts 210-211. Post-marketing reporting requirements and Food and Drug Administration withdrawal of approval are in §310.305 and §314.150, respectively. Similar requirements for premarket testing of chemical substances may be imposed by the Environmental Protection Agency under the Toxic Substances Control Act, 15 U.S.C. §§ 2601-71 (1982 and Suppl. 1992). See generally (Hanan, 1992).

2. In the case of prescription drugs, as post-marketing information accumulates, a drug will either remain a prescription drug, be released as an over-the-counter drug if safe enough, or pulled from the market if found to be more dangerous than initially believed. The idea behind the drug licensing scheme is to manage the uncertainty inherent in drug design and clinical trials both to ensure that efficacious drugs are supplied in a timely manner to sick people and to minimize the likelihood that there are unacceptable undiscovered risks of drug use.

3. A classic example (Fitts and Posner, 1967) involves two versions of a World War II aircraft, differing solely in the functions assigned to three key operating levers. Although aviators were instructed in the differences, they sometimes "forgot" under the strain of operations, acting as though they were flying a previously learned configuration. Lacking the influence needed to change the physical design of the aircraft, the human factors engineers compensated by placing tactually distinctive knobs on the levers.

REFERENCES

Aftermath of Chernobyl. 1986. Groundswell 9:1–7.
American Law Institute. 1965. Restatement (Second) of the Law of Torts.
Baum, A., R. Gatchel, and M. Schaeffer. 1983. Emotional, behavioral and physiological effects of chronic stress at Three Mile Island. Journal of Consulting and Clinical Psychology 12:349–359.
Berendes, H. W., and Y. J. Lee. 1993. Suspended judgment: The 1953 clinical trial of diethylstilbestrol during pregnancy: Could it have stopped DES use? Controlled Clinical Trials 14:179–182.
Beyth-Marom, R. 1982. How probable is probable? Journal of Forecasting 1:257–269.
Bostrom, A. 1990. A Mental Models Approach to Exploring Perceptions of Hazardous Processes. Ph.D. dissertation. School of Urban and Public Affairs, Carnegie Mellon University.
Bostrom, A., B. Fischhoff, and M. G. Morgan. 1992. Characterizing mental models of haz-

ardous processes: A methodology and an application to radon. Journal of Social Issues 48(4):85–100.

Brennan, T. A. 1988. Causal chains and statistical links: The role of scientific uncertainty in hazardous substance litigation. Cornell Law Review 73:469–533.

Carroll, J. M., and J. R. Olson. 1988. Mental models in human-computer interaction. Pp. 45–65 in Handbook of HumanComputer Interaction, M. Helander, ed. Amsterdam: Elsevier.

Chapanis, A. 1959. Research Methods in Human Engineering. Baltimore, Maryland: Johns Hopkins University Press.

Chi, M. T., P. J. Feltovich, and R. Glaser. 1981. Categorization and representation of physics problems by experts and novices. Cognitive Science 15:121–152.

Cohen, B., and I. S. Lee. 1979. A catalog of risks. Health Physics 36:707–722.

Covello, V. T., P. M. Sandman, and P. Slovic. 1988. Risk Communication, Risk Statistics, and Risk Comparisons: A Manual for Plant Managers. Washington, D.C.: Chemical Manufacturers Association.

Cranor, C. F. 1993. Regulating Toxic Substances: A Philosophy of Science and the Law. New York: Oxford University Press.

Crouch, E.A.C., and R. Wilson. 1981. Risk/Benefit Analysis. Cambridge, Mass.: Ballinger.

Eisenberg, T., and J. A. Henderson, Jr. 1993. Products liability cases on appeal: An empirical study. Justice System Journal 16:117–138.

Ericsson, K. A. 1988. Concurrent verbal reports on text comprehension: A review. Text 8(4):295–325.

Fischhoff, B. 1982. Debiasing. Pp. 422–444 in Judgment Under Uncertainty: Heuristics and Biases, D. Kahneman, P. Slovic, and A. Tversky, eds. New York: Cambridge University Press.

Fischhoff, B. 1987. Treating the public with risk communications: A public health perspective. Science, Technology, and Human Values 12:13–19.

Fischhoff, B. 1989. Eliciting knowledge for analytical representation. IEEE Transactions on Systems, Man and Cybernetics 13:448–461.

Fischhoff, B. 1991. Value elicitation: Is there anything in there? American Psychologist 46:835–847.

Fischhoff, B., A. Bostrom, and M. J. Quadrel. 1993. Risk perception and communication. Annual Review of Public Health 14:183–203.

Fischhoff, B., S. Lichtenstein, P. Slovic, S. L. Derby, and R. L. Keeney. 1981. Acceptable Risk. New York: Cambridge University Press.

Fischhoff, B., and D. MacGregor. 1983. Judged lethality: How much people seem to know depends upon how they are asked. Risk Analysis 3:229–236.

Fischhoff, B., P. Slovic, and S. Lichtenstein. 1977. Knowing with certainty: The appropriateness of extreme confidence. Journal of Experimental Psychology: Human Perception and Performance 20:159–183.

Fischhoff, B., P. Slovic, and S. Lichtenstein. 1978. Fault trees: Sensitivity of assessed failure probabilities to problem representation. Journal of Experimental Psychology: Human Perception and Performance 4:330–344.

Fischhoff, B., P. Slovic, and S. Lichtenstein. 1980. Knowing what you want: Measuring labile values. Pp. 117–141 in Cognitive Processes in Choice and Decision Behavior, T. Wallsten, ed. Hillsdale, N.J.: Lawrence Erlbaum Associates.

Fischhoff, B., P. Slovic, and S. Lichtenstein. 1983. The "public" vs. the "experts": Perceived vs. actual disagreement about the risks of nuclear power. Pp. 235–249 in Analysis of Actual vs. Perceived Risks, V. Covello, G. Flamm, J. Rodericks, and R. Tardiff, eds. New York: Plenum.

Fischhoff, B., and O. Svenson. 1988. Perceived risk of radionuclides: Understanding public understanding. Pp. 453–471 in Radionuclides in the Food Chain, J. H. Harley, G. D. Schmidt, and G. Silini, eds. Berlin: Springer-Verlag.

Fischhoff, B., S. Watson, and C. Hope. 1984. Defining risk. Policy Sciences 17:123–139.

Fitts, P. M., and M. I. Posner. 1967. Human Performance. Belmont, Calif.: Brooks/Cole.

Garnham, A. 1987. Mental Models as Representations of Discourse and Text. New York: Halsted Press.

Gilovich, T. 1993. How We Know What Isn't So. New York: Free Press.

Grimes, D. A. 1993. Technology follies: The uncritical acceptance of medical innovation. Journal of the American Medical Association 269:3030–3033.

Hanan, A. 1992. Pushing the environmental regulatory focus a step back: Controlling the introduction of new chemicals under the Toxic Substances Control Act. American Journal of Law and Medicine 18:395–421.

Hasher, L., and R. T. Zacks. 1984. Automatic and effortful processes in memory. Journal of Experimental Psychology: General 108:356–388.

Henderson, J. A., Jr. 1973. Judicial review of manufacturer's conscious design choices: The limits of adjudication. Columbia Law Review 73:1531–1578.

Henderson, J. A., Jr. 1979. Renewed judicial controversy over defective design: Toward the preservation of an emerging consensus. Minnesota Law Review 63:773–807.

Henderson, J. A., Jr. 1991. Judicial reliance on public policy: An empirical analysis of products liability decisions. George Washington Law Review 59:1570–1613.

Henderson, J. A., Jr., and A. D. Twerski. 1990. Doctrinal collapse in products liability: The empty shell of failure to warn. New York University Law Review 65:265–327.

Henrion, M., and B. Fischhoff. 1986. Assessing uncertainty in physical constants. American Journal of Physics 54:791–798.

Hensler, D. R., M. S. Marquis, A. F. Abrahamse, S. H. Berry, P. A. Ebener, E. G. Lewis, E. A. Lind, R. J. MacCoun, W. G. Manning, J. A. Rogowski, and M. E. Vaiana. 1991. Compensation for Accidental Injuries in the United States. Santa Monica, Calif.: RAND.

Hohenemser, C., M. Deicher, A. Ernst, H. Hofsass, G. Lindner, and E. Recknagel. 1986. Chernobyl: An early report. Environment 28:6–43.

Huber, P. W., and R. E. Litan, eds. 1991. The Liability Maze: The Impact of Liability Law on Safety and Innovation. Washington D.C.: Brookings Institution.

Hynes, M., and E. Vanmarcke. 1976. Reliability of Embankment Performance Predictions. Proceedings of the ASCE Engineering Mechanics Division Specialty Conference. Waterloo, Ontario: University of Waterloo Press.

Institute of Medicine. 1986. Confronting AIDS: Directions for Public Health, Health Care, and Research. Washington, D.C.: National Academy Press.

Johnson, E. J., and A. Tversky. 1983. Affect, generalization and the perception of risk. Journal of Personality and Social Psychology 45:20–31.

Jungermann, H., R. Schutz, and M. Thuring. 1988. Mental models in risk assessment: Informing people about drugs. Risk Analysis 8:147–159.

Kahneman, D., P. Slovic, and A. Tversky, eds. 1982. Judgments Under Uncertainty: Heuristics and Biases. New York: Cambridge University Press.

Kempton, W. 1991. Lay perspectives on global climate change. Global Environmental Change 6:183–208.

Kintsch, W. 1986. Learning from text. Cognition and Instruction 3:87–108.

Krimsky, S., and A. Plough. 1988. Environmental Hazards. Dover, Mass.: Auburn House.

Krohn, W., and P. Weingart. 1987. Commentary: Nuclear power as a social experiment— European political "fall out" from the Chernobyl meltdown. Science, Technology, & Human Values 12:52–58.

Krzysztofowicz, R. 1983. Why should a forecaster and a decision maker use Bayes Theorem? Water Resources Research 19:327–336?

Lamken, J. A. 1989. Note, efficient accident prevention as a continuing obligation: The duty to recall defective products. Stanford Law Review 42:103–162.

Laughery, K. R. 1993. Everybody know—or do they? Ergonomics in Design (July):8–13.

Leventhal, H., and L. Cameron. 1987. Behavioral theories and the problem of compliance. Patient Education and Counseling 10:117–138.

Lichtenstein, S., P. Slovic, B. Fischhoff, M. Layman, and B. Combs. 1978. Judged frequency of lethal events. Journal of Experimental Psychology: Human Learning and Memory 4:551–578.

Localio, A. R., A. G. Lawthers, T. A. Brennan, N. M. Laird, L. E. Hebert, L. M. Peterson, J. P. Newhouse, P. C. Weiler, and H. H. Hiatt. 1991. Relation between malpractice claims and adverse events due to negligence. New England Journal of Medicine 325:245–251.

MacGregor, D. G. 1989. Inferences about product risks: A mental modeling approach to evaluating warnings. Journal of Products Liability 12:75–91.

MacLean, D. 1987. Understanding the nuclear power controversy. In Scientific Controversies: Case Studies in the Resolution and Closure of Disputes in Science and Technology, H. T. Engelhadt, Jr. and A. L. Caplan, eds. Cambridge: Cambridge University Press.

Magat, W. A., and W. K. Viscusi. 1992. Informational Approaches to Regulation. Cambridge, Mass.: MIT Press.

Mahoney, M. J. 1979. Psychology of the scientist: An evaluative review. Social Studies of Science 9:349–375.

Mazur, A. 1973. Disputes between experts. Minerva 11:243–262.

Merz, J. F. 1991a. An empirical analysis of the medical informed consent doctrine: Search for a "standard" of disclosure. Risk 2:27–76.

Merz, J. F. 1991b. Toward a Standard of Disclosure for Medical Informed Consent: Development and Demonstration of a Decision-Analytic Methodology. Ph.D. dissertation. Carnegie Mellon University.

Merz, J. F., B. Fischhoff, D. J. Mazur, and P. S. Fischbeck. 1993. A decision-analytic approach to developing standards of disclosure for medical informed consent. Journal of Products and Toxics Liabilities 15:191–215.

Morgan, M. G., B. Fischhoff, A. Bostrom, L. Lave, and C. J. Atman. 1992. Communicating risk to the public. Environmental Science and Technology 26:2048–2056.

Morgan, M. G., and M. Henrion. 1991. Uncertainty. New York: Cambridge University Press.

Murphy, A. H., and B. G. Brown. 1983. Forecast terminology: Composition and interpretation of public weather forecasts. Bulletin of the American Meteorological Society 65:13–22.

Murphy, A. H., S. Lichtenstein, B. Fischhoff, and R. L. Winkler. 1980. Misinterpretations of precipitation probability forecasts. Bulletin of the American Meteorological Society 61:695–701.

Murphy, A., and R. Winkler. 1984. Probability of precipitation forecasts: A review. Journal of the American Statistical Association 79:391–400.

National Research Council. 1989. Improving Risk Communication. Washington, D.C.: National Academy Press.

National Research Council. 1993. Workload Transition: Implications for Individual and Team Performance, B. M. Huey and C. D. Wickens, eds. Washington, D.C.: National Academy Press.

Nelkin, D., ed. 1978. Controversy: Politics of Technical Decisions. Beverly Hills, Calif.: Sage.

Nelkin, D. 1984. Science in the Streets: Report of the Twentieth Century Fund Task Force on the Communication of Scientific Risk. New York: Priority Press.

Nisbett, R. E., and L. Ross. 1980. Human Inference: Strategies and Shortcomings of Social Judgment. Englewood Cliffs, N.J.: Prentice Hall.

Norman, D. A. 1988. The Psychology of Everyday Things. New York: Basic Books.

Peterson, C. R., and L. R. Beach. 1967. Man as an intuitive statistician. Psychological Bulletin 69:29–46.

Poulton, E. C. 1968. The new psychophysics: Six models for magnitude estimation. Psychological Bulletin 69:1–19.

Poulton, E. C. 1982. Biases in quantitative judgments. Applied Ergonomics 13:31–42.

Quadrel, M. J. 1990. Elicitation of Adolescents' Risk Perceptions: Qualitative and Quantitative Dimensions. Ph.D. dissertation. Carnegie Mellon University.

Reason, J. 1990. Human Error. New York: Cambridge University Press.

Reder, L. M. 1985. Techniques available to author, teacher, and reader to improve retention of main ideas of a chapter. Pp. 37–64 in Thinking and Learning Skills: Research and Open Questions, Vol. 2, S. F. Chipman, J. W. Segal, and R. Glaser, eds. Hillsdale, N.J.: Lawrence Erlbaum Associates.

Richardson, D., J. Sorensen, and E. J. Soderstrom. 1987. Explaining the social and psychological impacts of a nuclear power plant accident. Journal of Applied Social Psychology 17:16–36.

Rohrmann, B. 1990. Analyzing and evaluating the effectiveness of risk communication programs. Unpublished manuscript. University of Mannheim.

Roth, E., G. Morgan, B. Fischhoff, L. Lave, and A. Bostrom. 1990. What do we know about making risk comparisons? Risk Analysis 10:375–387.

Rothman, S., and S. R. Lichter. 1987. Elite ideology and risk perception in nuclear energy policy. American Political Science Review 81:383–404.

Rustad, M. 1992. In defense of punitive damages in products liability: Testing tort anecdotes with empirical data. Iowa Law Review 77.

Saks, M. J. 1992. Do we really know anything about the behavior of the tort liability system and why not? University of Pennsylvania Law Review 140:1147.

Schriver, K. A. 1989. Plain Language for Expert or Lay Audiences: Designing Text Using Protocol-Aided Revision. Communications Design Center, Carnegie Mellon University.

Shlyakhter, A., D. Kammen, C. L. Broido, and R. Wilson. 1994. Quantifying the credibility of energy projections from trends in past data. Energy Policy 22:119–131.

Slovic, P. 1987. Perceptions of risk. Science 236:280–285.

Slovic, P., and B. Fischhoff. 1983. Targeting risks: Comments on Wilde's "Theory of Risk Homeostasis." Risk Analysis 2:227–234.

Slovic, P., B. Fischhoff, and S. Lichtenstein. 1979. Rating the risks. Environment 21:14–20, 36–39.

Slovic, P., S. Lichtenstein, and B. Fischhoff. 1984. Modeling the societal impact of fatal accidents. Management Science 30:464–474.

Svenson, O. 1981. Are we all less risky and more skillful than our fellow drivers? Acta Psychologica 47:143–148.

Turner, C. F., and E. Martin, eds. 1984. Surveying subjective phenomena. 2 Volumes. New York: Russell Sage Foundation.

Tversky, A., and D. Kahneman. 1973. Availability: A heuristic for judging frequency and probability. Cognitive Psychology 5:207–232.

Tversky, A., and D. Kahneman. 1981. The framing of decisions and the psychology of choice. Science 211:453–458.

Twerski, A. D., A. S. Weinstein, W. A. Donaher, and H. R. Piehler. 1980. Shifting perspectives in products liability: From quality to process standards. New York University Law Review 55:347–384.

Viscusi, W. K. 1992. Smoking. New York: Oxford University Press.

Vlek, C. A. J., and P. J. Stallen. 1980. Rational and personal aspects of risk. Acta Psychologica 45:273–300.

von Winterfeldt, D., and W. Edwards. 1986. Decision Analysis and Behavioral Research. New York: Cambridge University Press.

Voss, J. F., S. W. Tyler, and L. A. Yengo. 1983. Individual differences in the solving of social science problems. Individual Differences in Cognition 1:205–232.

Wade, J. W. 1973. On the nature of strict tort liability for products. Mississippi Law Journal 44:825–851.

Wallsten, T., and D. Budescu. 1983. Encoding subjective probabilities: A psychological and psychometric review. Management Science 29:135–140.

Wallsten, T., D. V. Budescu, A. Rapoport, R. Zwick, and B. Forsyth. 1986. Measuring the vague meanings of probability terms. Journal of Experimental Psychology: General 115:348–365.

Weinstein, A. S., A. D. Twerski, H. R. Piehler, and W. A. Donaher. 1978. Products Liability and the Reasonably Safe Product: A Guide for Management, Design, and Marketing. New York: Wiley & Sons.

Weinstein, N., ed. 1987. Taking Care: Understanding and Encouraging Self-Protective Behavior. New York: Cambridge University Press.

Wilde, G. J. S. 1982. The theory of risk homeostasis: Implications for safety and health. Risk Analysis 2:209–225.

Wilson, R. 1979. Analyzing the risks of everyday life. Technology Review 81(4):40–46.

Contributors

CHARLES W. BABCOCK, JR., has been an attorney on the legal staff of General Motors Corporation since 1971, concentrating in product liability and product regulatory law. He has participated in joint projects with the American Medical Association and in environmental education in cooperation with the U.S. Environmental Protection Agency. Before joining GM, Mr. Babcock served as a captain in the U.S. Marine Corps, where he was a judge advocate and military judge, and from 1969 to 1971 as an associate with a Kansas City law firm. Mr. Babcock has had numerous legal articles published in professional journals. He received his A.B. degree from the University of Missouri and his J.D. degree from Harvard University.

FRANÇOIS J. CASTAING is vice president for vehicle engineering for Chrysler Corporation. He is responsible for the design, development, and implementation of all vehicle engineering programs and the development of vehicle technology, including electric and alternative fuel vehicles. Before joining Chrysler, Mr. Castaing was with American Motors Corporation. He served as Renault USA product engineering director supporting the launch of the Renault Alliance in 1980 and worked as chief engineer for the Renault Gordini and Renault Sport. Mr. Castaing was born in Marseille, France, and is a graduate of Ecole Nationale Superieure d'Arts et Metiers.

PAUL CITRON is vice president of science and technology at Medtronic, Inc. He is responsible for corporate assessment and coordination of technology and for setting directions and priorities for corporate research.

Previous positions at Medtronic included serving as vice president of ventures technology and director, then vice president, of applied concepts research. Mr. Citron was elected a Founding Fellow of the American Institute of Medical and Biological Engineering in January 1993, has twice won the American College of Cardiology Governor's Award for Excellence, and in 1980 was inducted as a Fellow of the Bakken Society. He is the author of many publications and holds several pacing-related patents for medical devices. A member of the Institute of Electrical and Electronics Engineers, Mr. Citron was voted IEEE's Young Electrical Engineer of the Year in 1979. He received a B.S. degree in electrical engineering from Drexel University and an M.S. in electrical engineering from the University of Minnesota, where he was a research fellow in the Department of Neurology.

DENNIS R. CONNOLLY is a principal and a senior vice president of Johnson & Higgins, the world's largest privately held international insurance brokerage, human resource, and employee benefits consulting firm. Mr. Connolly came to J&H from the American Insurance Association, where he was responsible for developing and implementing policy positions and for supervising liability issues. He has served on numerous task forces and study groups, including two Joint Insurance Trade Association Task Forces on Major Exposures, the National Association of Manufacturers Product Liability Task Force, and the Keystone Center's Program on Compensation for Environmental Injuries. Mr. Connolly is a member of the American Bar Association and the American Law Institute, was an adviser to the American Law Institute Compensation and Liability for Product and Process Injuries Project, and is a vice chairman of the American Bar Association's Committee on Energy Resources Law, Tort and Insurance Practice Section. He is a graduate of Colby College in Waterville, Maine, and has a J.D. degree from the Brooklyn Law School.

BENJAMIN A. COSGROVE retired in 1993 as senior vice president for technical and government affairs for the Boeing Commercial Airplane Group (BCAG). In that position, he was responsible for all liaison with regulatory agencies in matters of design and technology and was BCAG's senior executive in safety matters. He also formerly served as senior vice president and general manager of BCAG's Engineering Division, responsible for all engineering functions, flight test engineering and operations, and government technical contacts. Mr. Cosgrove has been associated with almost all Boeing jet aircraft programs, including positions as director of engineering for the 707/727/737 Division, chief project engineer and director of engineering for the 767 program, and director of engineering for the Everett Division (747/767 programs). He is the recipient of numerous honors and awards, including the Wright Brothers Memorial Trophy, the Ed Wells Technical Management Award, and membership in the National Academy of Engineering. Dr. Cosgrove has a B.S. in aeronautical engineering from the University of Notre Dame and in 1993 was awarded an honorary doctorate in engineering by his alma mater.

BARUCH FISCHHOFF is professor of social and decision sciences and of engineering and public policy at Carnegie Mellon University. His current research includes risk communication, adolescent decision making, evaluation of environmental damages, and insurance-related behavior. He serves on the editorial boards of several journals and is recipient of several awards, including the American Psychological Association's Early Career Awards for Distinguished Scientific Contribution to Psychology (1980) and for Contributions to Psychology in the Public Interest (1991), and the Distinguished Achievement Award (1991) from the Society for Risk Analysis. Dr. Fischhoff is a member of the Institute of Medicine of the National Academy of Sciences. He received his B.S. in mathematics from Wayne State University and M.A. and Ph.D. degrees in psychology from the Hebrew University of Jerusalem.

PETER W. HUBER is a lawyer and writer. He is a senior fellow of the Manhattan Institute for Policy Research and serves as counsel to the Washington, D.C., law firm of Kellogg, Huber & Hansen. He clerked on the D.C. Circuit Court of Appeals for Judge Ruth Bader Ginsburg and then on the U.S. Supreme Court for Justice Sandra Day O'Connor. Dr. Huber's professional expertise is in liability law and safety regulation. He is the author of *Liability* (1988); *The Geodesic Network: 1987 Report on Competition in the Telephone Industry* (1987); *The Liability Maze* (1991); *Galileo's Revenge: Junk Science in the Courtroom* (1991); *Federal Telecommunications Law* (1992); and *The Geodesic Network II: 1993 Report on Competition in the Telephone Industry* (1992). He writes a regular column for *Forbes*, and his articles have appeared in journals, magazines, and many newspapers. Dr. Huber has a doctorate in mechanical engineering from the Massachusetts Institute of Technology, where he served as an assistant and later associate professor for six years, and holds a law degree from the Harvard Law School.

JANET HUNZIKER is a program officer at the National Academy of Engineering, where most of her work focuses on issues related to the management of technological innovation. She has also organized numerous projects in the international area. Ms. Hunziker has a B.S. from Concordia College and an M.B.A. from the University of Maryland.

R. WILLIAM IDE III is a partner in the law firm of Long, Aldridge &

Norman in Atlanta, Georgia. From September 1, 1993, through August 1994, he was president of the American Bar Association (ABA). Mr. Ide was law clerk to Judge Griffin Bell and adjunct professor at the Florida State University College of Law. In addition to his long-standing involvement with the ABA, Mr. Ide is the recipient of many professional honors and awards, including the Arthur Van Briesen Award of the National Legal Aid and Defender Association; Designated Amicus Curiae of the Supreme Court of Georgia for Contribution to Administration of Justice; and commendation from the state of Georgia for outstanding service as chair of the Georgia Criminal Justice Council. Mr. Ide has a B.A. from Washington and Lee University, a law degree from the University of Virginia, and an M.B.A. from Georgia State University.

MARVIN E. JAFFE retired in 1994 as president of the R.W. Johnson Pharmaceutical Research Institute, the research and development organization that supports Ortho Pharmaceutical, McNeil Pharmaceutical, Ortho Biotech, and Cilag, all members of the Johnson & Johnson family of companies. Dr. Jaffe began his industrial career at Merck and spent 18 years with that company, rising to the position of senior vice president, Medical Affairs. Dr. Jaffe is a fellow of the American Academy of Neurology and the Royal Society of Medicine in England, and he serves on the advisory committee to the Harvard-MIT Division of Health Sciences and Technology. He has published extensively in medical journals on his research in the field of cerebrovascular diseases and is an expert in cerebral metabolism and pharmacology. Dr. Jaffe graduated from Temple University and received his medical degree from Jefferson Medical College in Philadelphia.

TREVOR O. JONES is chairman and chief executive officer of International Development Corporation, management consultants. Mr. Jones retired as chairman of the board of Libbey-Owens-Ford Company in 1994 and previously held the positions of president and chief executive officer. Before joining Libbey-Owens-Ford in 1987, Mr. Jones held positions at TRW, Inc., where he was group vice president for sales, marketing, strategic planning, and business development activities for the Automotive Worldwide Sector, and at General Motors, where his last position was as director of General Motors Proving Grounds. While at GM, he also directed Delco's program of applying aerospace technology to automotive electronic and safety systems. Mr. Jones is a member of the National Academy of Engineering and a fellow of the British Institute of Electrical Engineers, the American Institute of Electrical and Electronics Engineers, and the Society of Automotive Engineers. He is a recipient of the U.S. Department of Transportation Safety Award for Engineering

Excellence and the H. H. Bliss Award from the Center for Responsive Law, both for his contributions to inflatable restraint systems development. He holds many patents and has lectured and authored numerous papers on the subjects of automotive electronics and occupant safety. Mr. Jones is a native of England, where he completed his formal engineering education in electrical and mechanical engineering.

ALEXANDER MacLACHLAN retired as senior vice president of DuPont Research and Development in 1994. He joined the company in the Engineering Department and subsequently held numerous positions, including leader of research groups in photo imaging and in DuPont de Nemours (Deutschland) GmbH, director of the Research and Development Division of the Chemicals and Pigments Department, assistant director, then director of the Central Research and Development Department, and senior vice president of technology. Dr. MacLachlan is a member of the Board of Overseers, Fermi National Laboratory; a former director of the Industrial Research Institute; and a member of the National Academy of Engineering. He has published numerous articles in technical journals and holds several patents. Dr. MacLachlan has a B.S. in chemistry from Tufts College and a Ph.D. in physical organic chemistry from the Massachusetts Institute of Technology.

JON F. MERZ is an associate policy analyst with the RAND Corporation. He has more than seven years of risk analysis experience, as well as more than three years of experience in commercial and intellectual property law. His primary research interests involve technological risks and social policy, risk assessment and communication, and individual, regulatory, and judicial decision making under uncertainty. Dr. Merz holds a B.S. in nuclear engineering from Rensselaer Polytechnic Institute, an M.B.A. from the University of North Florida, a J.D. from Duquesne University Law School, and a Ph.D. in engineering and public policy from Carnegie Mellon University.

RICHARD M. MORROW is retired chairman of the board and chief executive officer of Amoco Corporation. Mr. Morrow joined Amoco Production Company, the Amoco subsidiary engaged in domestic exploration and production of oil and natural gas, in 1948. During the next two decades he held a number of engineering and managerial positions at various company locations in the United States. In 1966, Mr. Morrow was named as an executive vice president of Amoco International Oil Company, the Amoco subsidiary in charge of all overseas oil operations, before being named executive vice president of Amoco Chemical Company in 1970 and president of Amoco Chemical in 1974. He is a director of numerous companies, a former chairman of the National Academy of Engineering, and a trustee of the University of Chicago and Rush-Presbyterian-St. Luke's Medical Center. Mr. Morrow has a B.S. in mining and petroleum engineering from Ohio State University.

BRUCE E. PETERMAN is senior vice president of aircraft development for Cessna Aircraft Company. A 40-year Cessna veteran, Mr. Peterman has served as a flight test engineer, chief of propulsion, manager of technical engineering, chief engineer, vice president of engineering, and senior vice president of operations. An associate fellow of the American Institute of Aeronautics and Astronautics, Mr. Peterman served as a member of the Aerospace Council and Technical Board of the Society of Automotive Engineers. He is a member of the Industry Advisory Committee for the National Institute for Aviation Research at Wichita State University and a life member of the WSU School of Engineering Dean's Circle, from whom he received the 1993 Distinguished Engineers Service Award. He has served on the advisory committees for Kansas University's Aerospace Engineering Department and School of Engineering, and as a trustee of the Kansas University Center for Research. He is on the board of directors of Kansas Technology Enterprise Corporation and the board of governors of Wichita State University Endowment Association. Mr. Peterman has an M.S. in aeronautical engineering from Wichita State University and is an instrument-rated, multiengine pilot.

SUSAN ROSE-ACKERMAN is Henry R. Luce Professor of Jurisprudence (Law and Political Science), Yale University, and codirector of the Center for Studies in Law, Economics and Public Policy, Yale Law School. Professor Rose-Ackerman is the author of *Rethinking the Progressive Agenda: The Reform of the American Regulatory State* (1992); (with Estelle James) *The Nonprofit Enterprise in Market Economies* (1986); *Corruption: A Study in Political Economy* (1978); and (with others) *The Uncertain Search for Environmental Quality* (1974). Her forthcoming book on comparative administrative law is entitled *Controlling Environmental Policy: The Limits of Public Law in Germany and the United States*. Professor Rose-Ackerman earned her bachelor's degree in economics from Wellesley College and her Ph.D. in economics from Yale University.

VICTOR E. SCHWARTZ is senior partner in the Washington, D.C., law firm of Crowell & Moring. He cochairs the firm's Torts and Insurance Practice Group. His practice involves litigation, the development of legislation, and product liability loss prevention. He is also an adjunct professor at Georgetown University Law Center. Following positions on the faculty and as acting dean of the University of Cincinnati College of Law, Mr.

Schwartz chaired the Federal Interagency Task Force on Product Liability and received the Secretary of Commerce's Special Medal for his efforts. He drafted the Uniform Product Liability Act, which has been the basis for most state legislation on product liability. Mr. Schwartz serves on the Advisory Committee of the Restatement of Torts (Third): Product Liability Project and chairs the Civil Justice Reform Committee of the American Legislative Exchange Council and the Legislative Committee of the ABA Litigation Section. He is a director of, and general counsel to, the American Tort Reform Association. In May 1994 the *National Law Journal* included Mr. Schwartz in its listing of the 100 most influential attorneys in the United States. Mr. Schwartz received his bachelor's degree from Boston University and his J.D. from Columbia University, where he was an editor of the *Columbia Law Review*.

FREDERICK B. SONTAG is president of Unison Industries, a $50 million manufacturer of aviation ignition systems and other engine components with plants in Jacksonville, Florida, and Rockford, Illinois. Unison ignitions provide sparks to run piston engine aircraft as well as the energy to light the fuel mixture on larger turboprop and turbojet aircraft. Almost every commercial airplane flying, from trainers to 747s, contains a Unison product. Mr. Sontag held various corporate positions before purchasing Slick Electro of Rockford, Illinois, in a leveraged buyout in 1980 and later renaming the company Unison. In 1989 Unison purchased the ignition product line from the Bendix Engine Products Division of Allied-Signal, Inc. and moved its headquarters to Jacksonville, Florida. Mr. Sontag is past chairman of the board of the General Aviation Manufacturers Association (GAMA), has served as GAMA's Product Liability Committee chairman, and is a trustee of Harvey Mudd College. He holds a B.S. in physics from Harvey Mudd College, an M.S. in physics from the University of Nevada, and an M.B.A. from Harvard Business School.

Index

T

Technical innovation
  in chemical materials, 47
  in commercial aviation, 114-116
  inherent risk in, 24-25, 49
  insurability of, 135
  in medical device manufacture, 54-55
  public evaluation of, 165-166
  speed of, 79
  vs. technical correction, 10, 79
  See also Obstacles to innovation; Safety innovation
Texas, 33, 34
Thalidomide, 120-121, 122
Three Mile Island, 170
Tort reform
  call for federal standards, 6-7, 127
  contract law in, 147-148
  damage award limits, 147
  engineers in reform of, 97, 101
  expanded regulation vs., 99-101
  express warranty issues in, 32
  federal initiatives, 34
  general aviation industry and, 76
  incentive-based regulatory statutes, 155-156, 157
  insurance and, 137
  international competition and, 34-35
  meaning of, 111 n.61
  need for, 6, 151, 157
  pharmaceutical industry, 127

prospects for, 97, 102-103
public understanding, 2
recommendations, 7, 41, 52
regulatory reform vs., 126, 148
research needs, 42
scientific evidence, 14, 17, 127
social insurance system and, 104-105, 127
state initiatives, 33-34, 102-103
statute of limitations, 34
statutory reform vs., 152-154
TXO Production Corporation v. Alliance Resources, 88

U

University research, 65

V

Vaccines, 61, 136, 140
Viscusi, W. Kip, 84, 99

W

Waller, Patricia, 92-93
Warranty cards, 32-33
Whooping cough, 140
Workplace injuries/fatalities, 23, 100
  baseline risk standards, 153
  express warranty issues in, 31

Z

Zero-risk, 5, 27